TOGETHER IN ORBIT

The Origins of International Participation in the Space Station

John M. Logsdon

NASA History Division
Office of Policy and Plans
NASA Headquarters
Washington, DC 20546

Monographs in Aerospace History #11
November 1998

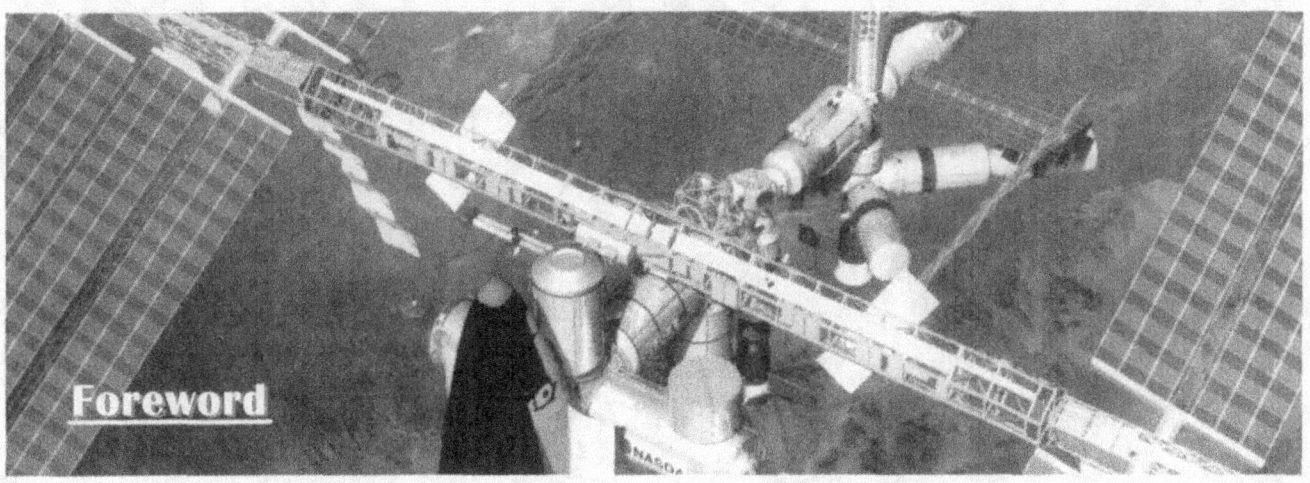

Foreword

FROM VIRTUALLY THE BEGINNING of the twentieth century, those interested in the human exploration of space have viewed as central to that endeavor the building of an Earth-orbital space station that would serve as the jumping-off point to the Moon and the planets. Always, space exploration supporters believed, a permanently occupied space station was a necessary outpost in the new frontier of space. The more technically minded recognized that once humans had achieved Earth orbit about 250 miles above the surface—the presumed location of any space station—the vast majority of the atmosphere and the gravity well would have been conquered, and then human beings were about halfway to anywhere they might want to go.

Space station advocates also recognized that the scientific and technological challenge of building an Earth-orbital space station was daunting and that pooling the resources of many of the spacefaring nations of the world would maximize the probabilities of success. Thus, when the space station project was born in the mid-1980s, it almost immediately became an international program. This monograph describes the very early process of conceptualizing the international partnership and crafting its contours during the period between 1984 and 1988.

This study was completed by John M. Logsdon of George Washington University in late 1991, but it was not published in a form suitable for wide circulation. With the incorporation of the Russian Federation into the space station partnership, it occurred to Dr. Logsdon that a full account of the origins of international involvement in the space station program might be of interest to many people, particularly as the initial launches of space station elements draw near and the process of assembling and then beginning to use the International Space Station is imminent.

Logsdon has made revisions to the text as it stood in 1991, adding a short concluding analysis that brings the study to the present, inserting recent publications into the footnotes, and fixing a few grammatical or linguistic infelicities. It seemed especially appropriate to recognize briefly that the partnership begun in 1984 and described in this account had been augmented by the 1993 invitation to Russia from the original partners to join them in the station enterprise, but other than that Logsdon let the study stand.

This is not the full story of the international partnership that has worked to build the International Space Station. Indeed, such a history would require much additional research. Nor does it contain any significant detail on the Russian partnership. Only the passage of time coupled with thoughtful reflection will make such a history possible. We are presently working toward a full-fledged history of the international partnership.

This is the eleventh in a series of monographs prepared under the auspices of the NASA History Division. The **Monographs in Aerospace History** series is designed to make available a wide variety of investigations relative to the history of aeronautics and space. These publications are intended to be tightly focused in terms of subject, relatively short in length, and reproduced in an inexpensive format to allow timely and broad dissemination to researchers in aerospace history. Suggestions for additional publications in the **Monographs in Aerospace History** series are welcome.

Roger D. Launius
NASA Chief Historian
September 15, 1998

Table of Contents

Acknowledgments

THIS STUDY TOOK a long time to complete, for a variety of justifiable and not so reasons. Along the way, the number of people who have helped has become very large, and I am sure that I will fail to give due credit and thanks to all who deserve it.

The study was carried out under contract to the NASA History Division, using funds provided by the Office of Space Station. Robert Freitag and Terence Finn, of the latter office, recognized the historical significance of the space station program and were willing to support outside, independent scholars to track the evolution of the program in near to real time. They also supported the creation of a Space Station Historical Archive; its manager, Adam Gruen, and his assistants did a superlative job of assembling documents and other materials from the early days of the program, from which I have drawn extensively in preparing this study. Freitag and Finn also read several drafts of the study and provided their on-the-spot perspectives on how events unfolded. Overseeing the space station historical effort and this study until she moved to the Administrator's office was NASA Historian Sylvia Fries, who provided gentle but firm guidance and insightful comments on early drafts of the study. Roger Launius was Sylvia's successor as Director of the History Division, and he was understanding as I pushed to finish the study.

Of the many others within NASA who helped me locate documents, provided essential corrective comments as I went along, and encouraged me to get the job done, I owe particular thanks to Peggy Finarelli. She trusted me enough to provide access to material not often available to an outside scholar, and then worked with me to make sure that I had not inadvertently violated her trust. Dick Barnes provided extensive insights and comments and, with Peggy, was instrumental in opening NASA files for my use. Al Condes worked with me on document access, and Lyn Wigbels and Ken Pedersen read the manuscript and provided helpful comments.

Individuals from U.S. station partners who went out of their way to be helpful include George van Reeth, Ian Pryke, Gabriel Lafferanderie, and Jean Arets of the European Space Agency; Mac Evans, Karl Doetsch, and Bill Cockburn of Canada; and Masanori Nagatomo, Shinichi Nakayama, and Yasahiro Kawasaki of Japan. Of course, the study would not have been possible without the willingness of many in the United States and overseas to take time for an interview.

At George Washington University, Henry Hitchcock provided valuable research assistance as he pushed to complete his own dissertation on the space station project. Lois Berdaus and Paul McDonnell typed early drafts of the study until I finally learned to use a word processor (and spell check!).

Underpinning this whole effort is an attitude within NASA that what the agency does is of lasting significance, is paid for by public funds, and should be open for scrutiny by outsiders such as me. Colleagues from abroad are amazed at the openness and accessibility of U.S. government officials and the willingness of government agencies to open all but their most sensitive files to external examination, if the purpose for doing so is valid. NASA has been a model within the government in this respect, at least as far as my experience is concerned.

When I prepared the manuscript for publication in this monograph series, Kerry Murray, a graduate research assistant at the Space Policy Institute, mastered the modern technology of a scanner so that there was no need to retype major portions of the manuscript, and she otherwise was of great assistance in getting the document ready for publication. She has my thanks.

I am grateful for the assistance offered by all those mentioned above and by others who contributed to this study. It goes without saying that I alone am responsible for all errors of fact and interpretation in this work and that the conclusions and findings are mine and do not necessarily reflect the views of the National Aeronautics and Space Administration or George Washington University.

The time when the space station experiment in international cooperation can be tested in practice is fast approaching. It certainly has been a long time in coming!

John M. Logsdon
March 1998

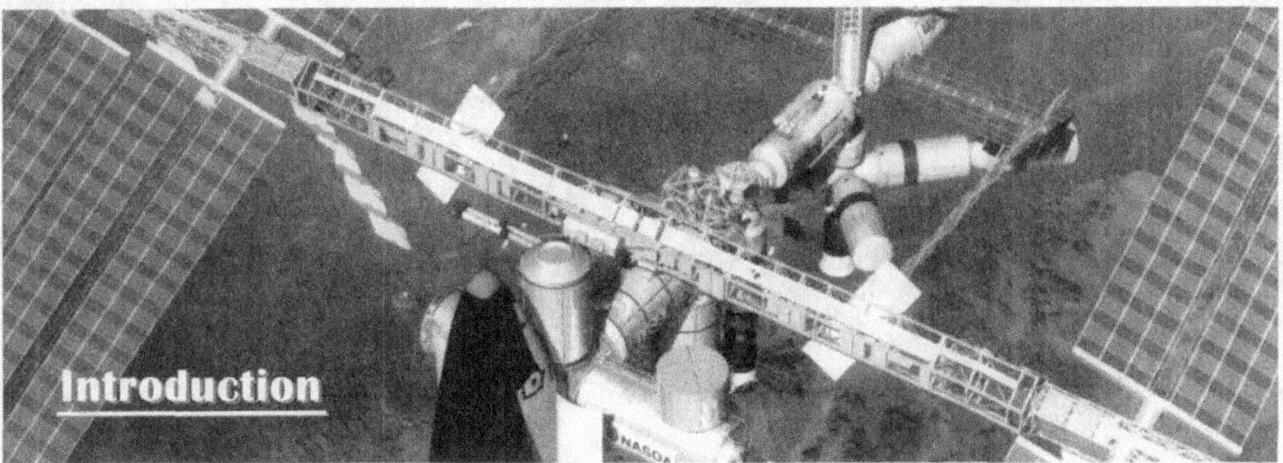

Introduction

On January 25, 1984, in his annual State of the Union address to a joint session of Congress, President Ronald Reagan announced that "tonight, I am directing NASA to develop a permanently manned space station and to do it within the decade." A few moments later, he added: "We want our friends to help us meet these challenges and share in their benefits. NASA will invite other countries to participate so we can strengthen peace, build prosperity, and expand freedom for all who share our goals."[1] Just over a year later, during the April–June 1985 period, Canada, Japan, and Europe accepted in principle the U.S. invitation to participate in the space station program. Thus was initiated the most extensive experiment in international technical cooperation ever undertaken.

This essay is a history and analysis of the steps leading to the origins of the space station partnership between the United States and its closest allies. It traces the process that led to the decision to invite other countries to participate in the project and their reasons for accepting that invitation. *Not* covered in this account are the difficult negotiations during the 1984–1988 period that led first to an initial set of agreements that allowed the prospective partners to work together during the early stages of the space station program and then to the final set of agreements creating the original space station partnership. Also, the 1993 invitation to the Russian Federation to join the original partners is not discussed, nor are the subsequent negotiations to revise the 1988 agreements.

International cooperation has been a hallmark of the U.S. civilian space program since its inception. It is fair to view that program not only as one pressing the frontiers of science and technology but also as an extremely important tool of U.S. propaganda; the cooperative aspects of the program were an important part of its propaganda aspects. The term *propaganda* has a somewhat negative connotation, but properly interpreted it means an attempt to project—to propagate—a positive message. The message sent to the world by the willingness of the United States to share the exploration of space with others is that of an open, dynamic, pioneering society, eager to share its capabilities and achievements with others. When that message was supplemented by the demonstration of technological and organizational skill that was Project Apollo, the space program clearly was a powerful means of validating the U.S. claim to world leadership.

Using the space program as an instrument of U.S. foreign policy was relatively easy when only the United States and the Soviet Union possessed the capability to put humans and their machines into orbit and beyond, particularly when the Soviet Union had a space program characterized by secrecy and by limited contact with countries other than its socialist allies. In the aftermath of the initial lunar landing, however, the United States recognized that other countries were seeking their own means of access to space and that the Cold War competition between the United States and the Soviet Union might be replaced by an era of détente. Faced with these emerging realities, during the 1969–1972 period, the United States consciously changed its approach to space cooperation from one that stressed data exchange, working together on scientific projects and providing launch services for the scientific satellites of other countries, to one that involved direct foreign participation in the human spaceflight program.

For the Soviet Union, this meant the highly symbolic Apollo-Soyuz Test Project that led to a 1975 "handshake in space." For traditional U.S. allies around the world, this meant an invitation to cooperate with the United States in the development of post-Apollo systems for human spaceflight.[2] For reasons described in detail in Chapter 2, this invitation resulted in Canada developing an essential hardware element for the Space Shuttle—the Remote Manipulator System—and Europe building a laboratory for use in the Shuttle's payload bay—Spacelab. These cooperative engagements were very different in character from any that had taken place before. They raised concerns about whether Europe and Canada had the technological capabilities to build sophisticated, highly reliable "human-rated" hardware, or whether the United States would have to provide them access to sensitive or proprietary technology for them to be successful in their projects. A contrary concern was whether the U.S. invitation would stimulate its partners to develop indigenous technological capabilities that then would be competitive with those possessed by the United States. Clearly, this was a form of cooperation qualitatively different from that involving a foreign scientist participating in an experiment flown aboard a U.S. spacecraft!

During the 1970s, both the United States and its partners went through a sometimes difficult period of learning to work together in developing new hardware for use by humans in space. The United States still was by far the dominant partner in the post-Apollo cooperative relationship. It was the United States that established the basic design of what would be developed and, more or less on a "take it or leave it" basis, told its potential partners what an acceptable contribution might be. As their confidence in their own capabilities increased during the 1970s, largely as the result of their success in the post-Apollo coopera-tion, this attitude was becoming unacceptable to Europe and Canada. If they were to be involved with the United States in future expensive and challenging hardware development projects, it would have to be on a more equitable basis.

This then was the context in which the United States took the initiative as the 1980s began to discuss with Europe, Canada, and Japan (which had not been able to participate in the post-Apollo program) possible cooperation in the "next logical step" in the development of space—the creation of a human outpost in Earth orbit, a space station. Once again, a major space undertaking was being put forth as a tool of U.S. policy—a policy that for more than two decades had used the space program to demonstrate what was best about American society. As the following pages suggest, embodying that objective and the recognition of U.S. leadership that accompanies it in a stable, harmonious space station partnership has proven no easy task.

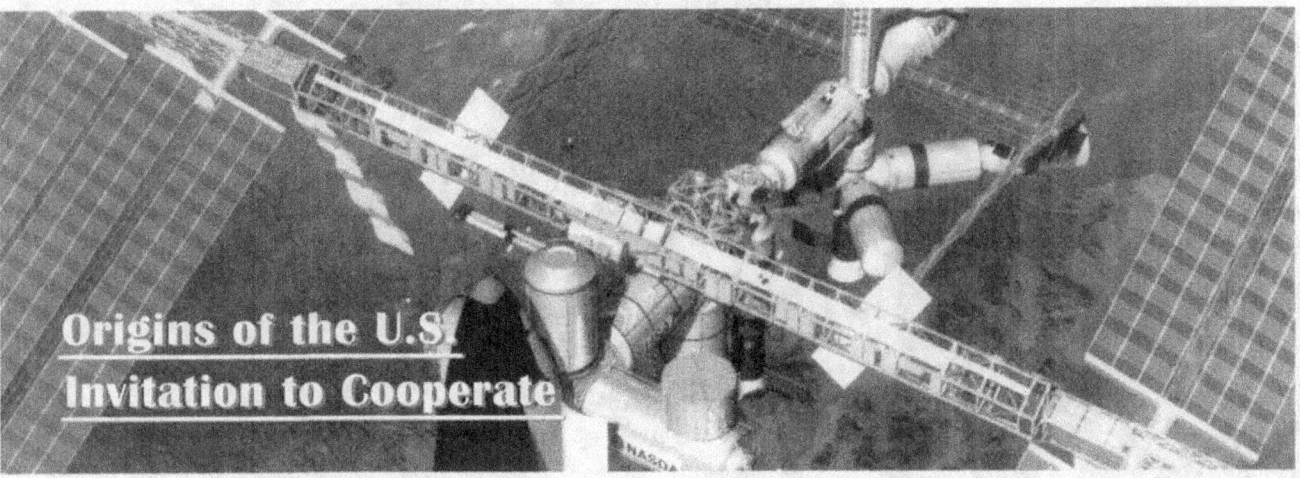

Origins of the U.S.
Invitation to Cooperate

Background

WHILE THE HIGHLY VISIBLE personal endorsement by President Reagan of foreign participation in the U.S. space station program may have come as a surprise to many in the United States and in potentially collaborating countries, the notion that the United States would welcome some form of international cooperation in the program certainly was not unexpected. During 1982 and 1983, as NASA had tried to lay the basis within the U.S. government for approval of its space station proposal, possible international involvement had been a subject of extensive discussion both within the United States and between the United States and its potential partners. That discussion itself built on a record of cooperation that extended back to the early years of the U.S. civilian space program in the late 1950s.[3]

The 1958 Space Act had set as one of NASA's objectives "cooperation by the United States with other nations and groups of nations."[4] NASA's cooperative activities were limited primarily to space science programs during the 1950s and 1960s, but as a post-Apollo program was being planned during the 1969–1971 period, there was a decision to broaden the basis of cooperation to include involvement in the development of hardware, particularly systems related to the human spaceflight program.[5]

NASA asked Europe, Canada, and Japan in late 1969 to consider ways of participating in its proposed post-Apollo program, which at that point was centered on an orbiting space station and a totally reusable launch vehicle called the Space Shuttle. Japan was just initiating its own general-purpose space agency (although it had had an active space science program for a decade). It took Japan some time to decide whether it wanted to respond to the U.S. invitation, particularly because its own space capabilities were at such an early stage of development. By the time that a response to the U.S. invitation was agreed on within Japan, the United States

had so changed the possibilities for international participation that there was no basis for Japanese involvement. Canada was eager to be involved. Several years of discussions led to an agreement that Canada would provide the Remote Manipulator System for the Space Shuttle. The Shuttle had turned out to be the only element of NASA's ambitious post-Apollo plans for human spaceflight that was approved by the Nixon administration.[6]

Negotiations between the United States and Europe on post-Apollo cooperation proved contentious and left many in Europe ultimately unsatisfied with the bargain struck.[7] Once it had been established, by 1971, that the Space Shuttle was the only major NASA project for the 1970s likely to receive funding, NASA and European space leaders agreed that Europe would examine three options for involvement in the Shuttle program:

1. Teaming between European and U.S. industry to develop specific parts of the Shuttle orbiter—for example, the tail and payload bay doors

2. European development of an orbital transfer vehicle, known as the "Tug," to move payloads from the Shuttle payload bay to other orbits

3. A Research and Applications Module, also called the "sortie can," to provide additional pressurized and unpressurized volume within the Shuttle payload bay for experimentation

During the 1971–1973 period, Europe spent approximately $20 million on studies of these alternatives; by 1972, Tug development had emerged as the preferred European contribution.

As the European studies progressed, however, the U.S. position with respect to the level and kind of European participation it wanted crystallized.

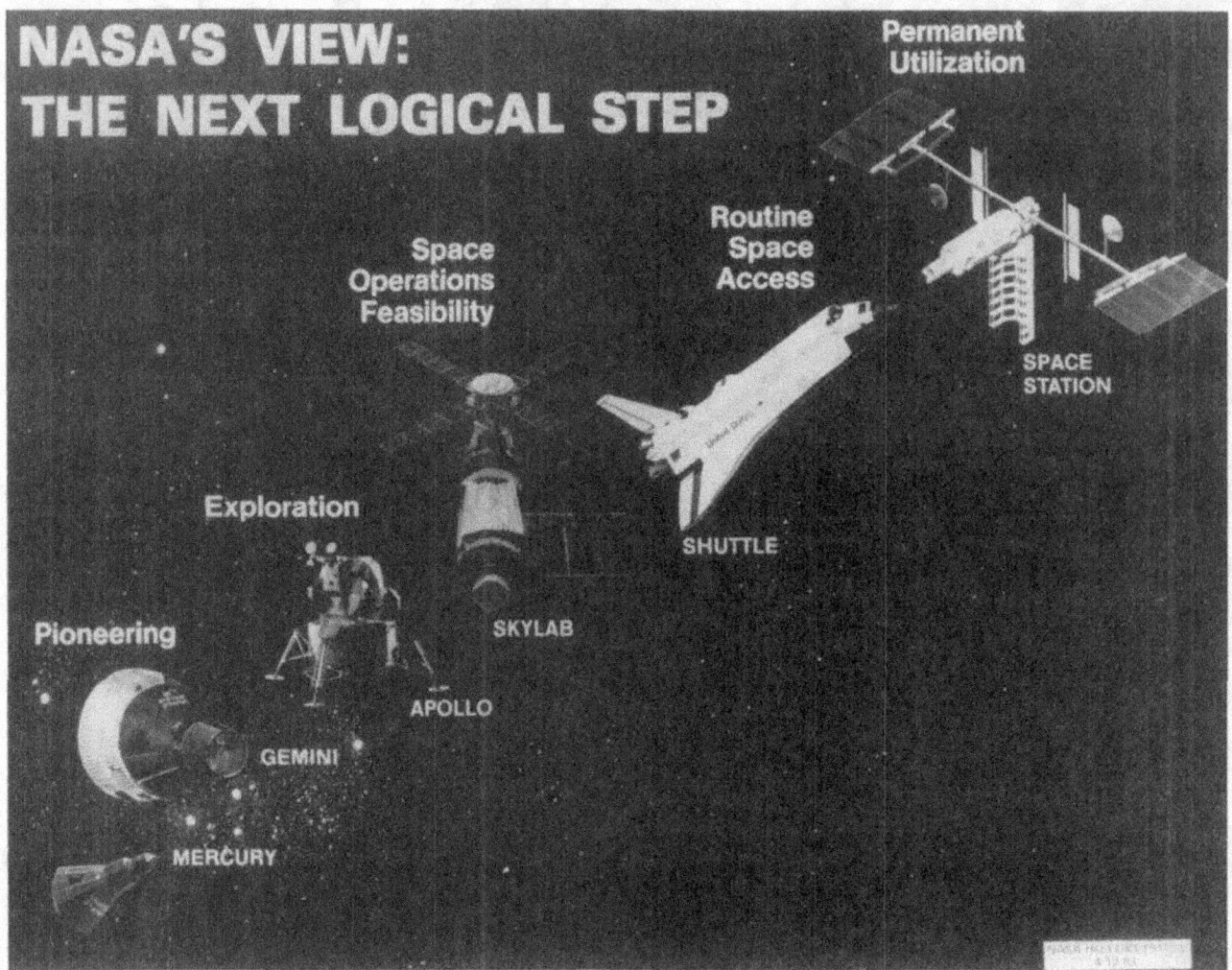

"NASA's View: The Next Logical Step"—In 1982, with the successful completion of the four Space Shuttle orbiter flight tests, NASA began planning activity to define a possible space station. The station was viewed as the next logical step in space. It built on the nation's past experience in space and provided, for the first time, the capability for permanent use of the space environment. (NASA photo 83-H-368).

First, the Nixon administration's interest in cooperating was later interpreted by the White House as an interest in European involvement in the use of space rather than in joint engineering projects. Second, NASA found that the European aerospace industry lagged approximately five to ten years behind U.S. industry. Therefore, NASA dropped the idea of joint development of technology, speculating that the United States might stand to lose more than it would gain. Third, NASA also decided that it did not want to depend on other countries for critical items on the Space Shuttle so that the Shuttle could fly independent of foreign activities. Fourth, NASA decided that, for safety reasons, it did not want to fly a Tug using liquid propellants, the only type Europe was studying. Moreover, there was real concern that Europe did not have all the technology to develop a Tug. Furthermore, the Tug was to be used to lift national security satellites to higher orbits, and the notion of

non-U.S. involvement with these highly classified satellites was not welcome to the national security community.

The U.S. government thus found itself in the position of having to walk back from the European perception of the cooperative possibilities in post-Apollo activities that had been encouraged by the way the United States and Europe had proceeded to define that cooperation.[8] By the time the U.S. position had been clarified in 1972, all Europe was offered, more or less on a "take it or leave it" basis, was the development of the Research and Applications Module. (The module was renamed Spacelab in 1973.)[9]

Some in Europe, particularly in France, were skeptical of the wisdom of close collaboration with the United States on expensive projects, preferring to concentrate on developing European capabilities for independent action. However, other countries, led by the Federal Republic of Germany and Italy, had become eager to become involved in developing hardware qualified for human spaceflight, thereby gaining skills in systems engineering and quality control. After hard negotiations within Europe, a "package deal" was agreed to in which funds were made available to develop a French-supported European launch vehicle (Ariane), to participate in the U.S. post-Apollo program through Spacelab development, and to develop a maritime communications satellite of primary interest to the United Kingdom. As part of the package deal, a new cooperative organization, the European Space Agency (ESA), was created to pool the technical and financial resources of European countries and to manage Ariane and Spacelab development and other cooperative projects.[10]

The U.S.-European agreement on the Spacelab project became a source of tension between the partners. At the time it was negotiated, NASA was projecting fifty or more Space Shuttle flights a year, at a cost per flight of less than $10 million; many of these flights were expected to require Spacelab use. Thus the thought was that about six sets of Spacelab hardware would be needed. Europe agreed in 1973 to develop the first Spacelab at its own expense and then transfer ownership to NASA; NASA agreed to purchase any additional Spacelabs required, with a minimum of one such purchase guaranteed.

By the time Spacelab was ready for use, its development costs had risen to almost $1 billion, rather than the approximately $250 million

This designer's conception shows some of the applications of an advanced Space Operations Center, which was studied by Boeing Aerospace Company for NASA. This advanced version of the "spaceport" shows the Space Shuttle unloading some of the modules that would comprise the system, including living and command control quarters; warehouses for food, water, and hydrazine; and service areas containing batteries and other necessary supplies. Other areas of this advanced concept include hangars for spacecraft, solar panels to provide power for the station, and construction equipment to handle large structures. The large structure containing several antenna reflectors is a communications platform that is about to be assembled to an Orbital Transfer Vehicle for a flight to a higher orbit in space. (NASA photo 81-H-793).

originally estimated. Projections of Space Shuttle usage had dramatically shrunk, and the United States decided to purchase only the one additional Spacelab it was obligated to buy, at a cost of $128 million. Any chance for Europe to recoup some of its development costs through Spacelab production thus vanished. The agreement provided for one joint U.S.-ESA Spacelab mission at no launch cost to ESA. After that, ESA would have to pay launch costs for any Spacelab missions it wanted to undertake. By the early 1980s, the combined costs of preparing the experiments for a Spacelab mission and paying Shuttle launch fees exceeded ESA's resources, and the agency was left in a position of not being able to afford the use of the system it had developed. (Germany undertook two Spacelab missions of its own—one in October 1985 and the second in April 1993.)

The U.S.-European interaction in the post-Apollo period has been described in some detail because it provided much of the context for U.S.-European discussions on potential space station cooperation.[11] In hindsight, some top European space officials described themselves as "stupid" in accepting the U.S. terms for involvement in its post-Apollo program, attributing their weakness to an early 1970s lack of confidence in European capabilities and to a belief that only through cooperation with the United States could those capabilities be improved.[12] Thus, according to this analysis, Europe was willing to pursue cooperation on almost any terms, no matter how one-sided. By contrast, in the early 1980s, with the completion of Spacelab and the successful development of the Ariane booster, Europe approached possible space station cooperation with a strong sense of its own capabilities and a determination to accept only an arrangement that recognized its position as a major spacefaring actor.

It is clear that Europe received substantial benefits from its post-Apollo cooperation with the United States. In particular, Europe gained experience in the systems-level management of complex space projects—an experience that was quickly applied to other European projects such as Ariane. The upgrading of Europe's management, technical, and human systems know-how obtained from the Spacelab experience was an important positive factor as the United States assessed possible international participation in the space station program.

Another positive byproduct of the post-Apollo cooperation between the United States and Europe and between the United States and Canada was a set of personal and organizational relationships biased toward continued cooperation. Those who had been most directly involved, by and large, found the experience programmatically productive and personally rewarding. Also, Canada successfully completed its contribution to the Space Shuttle and in the process earned the confidence of NASA engineers at the Johnson Space Center, some of whom were skeptical about the wisdom of non-U.S. involvement in America's human spaceflight efforts. Japan, forced to sit on the sidelines during Shuttle development, was determined not to be left out of the next major cooperative opportunity. As NASA began to explore the possibility of international involvement in the space station, there was a basis of positive experience and expectations among potential partners from which to proceed.

Laying the Foundation for International Cooperation[13]

The proposal to make a space station the central project in NASA's post-Apollo program had been decisively rejected by the Nixon administration during the 1969–1970 period. The concept that some kind of crewed orbital facility was an essential element of any plan for extensive space development did not die, however; during the 1970s, NASA sponsored a number of advanced studies of possible space station missions and configurations.[14] By early 1981, as the new administration of President Ronald Reagan took office, NASA's two major human spaceflight centers—the Johnson Space Center in Houston, Texas, and the Marshall Space Flight Center in Huntsville, Alabama—had each developed a preferred space station concept. The two concepts were very different in approach. The Marshall station began with a modest, human-tended platform that would gradually evolve into a permanently occupied facility; its primary mission was as a research laboratory. The Johnson concept was a large facility primarily intended to support space operations, such as in-orbit construction, fueling of spacecraft, and the preparation for human missions to the Moon and Mars. The two centers were traditional rivals within the decentralized NASA organization, and each was pushing NASA Headquarters to adopt its own approach to the agency's next major project.

Space Shuttle development was phasing down in 1981; the first flight of the Shuttle was scheduled for April. If NASA was to maintain its identity as an engineering organization responsible for developing large and complex hardware systems, particularly for human missions, it was clear that the agency needed to get a new post-Shuttle project approved soon.

It was in this context that the Reagan administration choices as NASA Administrator and Deputy

Administrator, James Beggs and Hans Mark, respectively, appeared before a Senate confirmation hearing on June 17, 1981. Beggs had served briefly in NASA in the late 1960s and then had become under secretary of the Department of Transportation; during the late 1970s, he had risen to a senior position with General Dynamics, a major aerospace corporation. Mark had been director of NASA's Ames Research Center in the early 1970s and had served as under secretary and then secretary of the Air Force during the Carter administration. Both were intimately familiar with space policy and program issues. They had actually been selected for their NASA positions in mid-March, and shortly thereafter Beggs had obtained Mark's agreement that "we would try to persuade the new administration to adopt the construction of a permanently manned orbiting space station as the next major goal in space."[15] Beggs and Mark announced that intent to the senators at their confirmation hearing.

It would take some time to develop the case for the space station and to convince Ronald Reagan to approve the project.[16] Before they could concentrate on station advocacy, Beggs and Mark had to bring the Space Shuttle into what could be characterized as operational status. They also had to fend off, as best they could, 1981 attempts by the new director of the Office of Management and Budget (OMB) to make major cuts in NASA's existing budget.[17] Thus, even though the two top NASA officials had publicly strongly endorsed the station as "the next logical step" in space, the station program took some time to pick up momentum, although early planning activities began almost immediately. An initial Space Station Conference was scheduled for November 1981 to inform individuals throughout NASA and the U.S. government of NASA's thinking to date and to lay the basis for the more intensive planning effort that all knew was required.

From the start, the possibility of international involvement in any station program that might be proposed was part of that planning. As mentioned earlier, a bias toward international involvement in its activities had been part of the NASA culture since the organization's inception. To those in charge of planning for the space station program, it was inconceivable that the United States would go forward with a major effort in space and not include some form of international cooperation. Typical of those who held this perspective was Robert Freitag, Deputy Director of the Advanced Program Office in the Office of Manned Space Flight, who had been one of the primary architects of NASA-European post-Apollo cooperation and saw international cooperation in the

space station, particularly with Europe, as a productive continuation of the relationship established during the 1970s.

Kenneth Pedersen, the Director of NASA's Office of International Affairs, was another advocate of international cooperation. Unlike Freitag, Pedersen was not a long-time NASA employee; he had come to the space agency in 1979 from his position as head of policy analysis and evaluation at the Nuclear Regulatory Commission. Pedersen's position, as the policy-level advisor on international affairs to the NASA Administrator and as NASA representative in discussions of international space matters with the White House and other executive branch agencies, gave him and his staff a different perspective than that held by people such as Freitag, who was working on international programmatic and technical issues in one of the line offices of the agency. While Freitag and his associates were enthusiastic advocates of cooperation within and outside NASA, Pedersen had to take a more cautious approach. He was fully aware of the skepticism about the benefit-risk ratio of large-scale international technological interactions that was widespread among key members of both the career national security community and the new Reagan administration.

If there was skepticism and even opposition within the space agency about the value of international involvement in NASA's major programs, it resided primarily in the field center people who had to deal with the added managerial complexity inevitably introduced by such involvement. While many at the Marshall Space Flight Center who had been involved with cooperation in the Spacelab program were receptive to international involvement in the space station, staff at the Johnson Space Center tended to be more dubious about the wisdom of intimate international partnerships.

When NASA convened the initial agencywide workshop on space station planning in November 1981, international involvement was a prominent agenda item, and the report of the workshop noted that:

There appears to be substantial foreign interest in NASA's future plans for its manned space activities. In some cases, this interest derives from existing contributions to NASA's Space Transportation System [STS]. Extending this cooperation by participating in a NASA space station seems a logical step to some countries and space agencies.

NASA can derive significant benefits from international participation in its programs if they are properly structured and controlled. These benefits may include economic cost sharing, access to unique or otherwise valuable expertise, and improvements in the linking of foreign programs to STS utilization.

The subject of potential international participation in a U.S. space station program must be approached carefully and proceed under clear assumptions and guidelines. A fundamental ground rule should be that planning for a space station will be conducted as if the entire project is to be developed as a wholly U.S. effort. Planning should proceed, however, on the basis that it does not foreclose international cooperation. Potential foreign participants should be encouraged to fund and undertake parallel studies of space station requirements and concepts which could benefit NASA in its design of the space station. Procedures should be developed to facilitate controlled exchanges of study results. All potential partners should be clearly informed that such exchanges during Phase A do not represent a commitment on the part of NASA to foreign involvement in the actual development of the station.[18]

These 1981 perspectives guided NASA's approach to possible international involvement in the station over the subsequent several years. Indeed, those within NASA responsible for technical-level liaison with Europe for some time had been discussing with their European colleagues the possibility of a U.S. space station program and of European involvement in it.[19] The approach articulated at the November conference reflected those discussions.

Another agenda item at the workshop was potential Department of Defense (DOD) involvement in the space station. The support of the national security community had been essential to gaining White House approval for the Space Shuttle, but a fair degree of tension in the NASA-DOD relationship had risen in the decade since. However, the workshop report noted that "the climate for initiating major new NASA/DOD space endeavors is improving." The report also recognized that "national security interests may have considerable impact on the feasibility or nature of international participation in a Space Station program."[20] NASA hoped to find a way to reconcile both DOD involvement and international participation, and thereby keep two influential constituencies involved as it attempted to gain political support for its plans.

Once a general approach to international involvement in station planning had been developed, the next step was to inform potential partners what it was.

For this purpose, NASA's international affairs chief Kenneth Pedersen convened a meeting at the Johnson Space Center on January 13, 1982. Pedersen called this meeting on his own authority, although he informed NASA Administrator Beggs that he was doing so. Pedersen had been one of the first senior NASA staff members to work closely with Beggs after he had been selected to head NASA. Beggs attended the Paris Air Show in June 1981 as NASA Administrator-designate, and he and Pedersen met with representatives of other countries to discuss NASA's future. These meetings and frequent one-on-one discussions made it clear to Pedersen that Beggs was an internationalist in orientation and, based on his experience with international cooperation and co-production of the General Dynamics F-16 fighter, understood the value to the United States of involving allies in major U.S. programs.[21] While there had been no formal decision by Beggs to begin the process of soliciting international participation in the space station, Pedersen in early 1982 felt on safe grounds in calling together representatives of potential partners from Europe, Canada, and Japan for a status report on space station planning and a discussion of the approach that NASA would take to assessing potential international involvement.

NASA's international partners during the preceding two decades had been critical of the organization for deciding by itself on the objectives and design of projects and only then inviting foreign involvement, on terms largely dictated by NASA. Pedersen's major point at the January meeting was that there would be a shift in NASA's approach; potential partners were being invited to become involved at a very early stage in program definition, so that their inputs could help influence NASA's choices and they could understand from the start options for their participation. This approach, he stressed, had risks as well as benefits. Pedersen told the foreign representatives at this and subsequent meetings: "[T]his is going to be for you an exciting and a frustrating process: exciting because I think you will see just how a program like this gets put together from the nuts and bolts stage; frustrating because you're going to suffer the stops and starts and uncertainties that all programs like this go through in the early stages."[22] As long as this situation was understood, said Pedersen, NASA was eager, under the guidelines articulated at the November 1981 meeting, to have its foreign counterparts begin to study possible ways of becoming involved as NASA's station plans took form.[23]

Not only its potential partners, but also NASA, were taking some risks in this new approach. Since its inception, NASA had structured its international cooperative programs under a set of guidelines that provided the agency almost total control over the character of those activities. Key to those guidelines were the notions of cooperative projects being undertaken only when they were clearly of mutual interest, no exchange of funds or unwarranted transfer of technology, "clean" technological interfaces, and NASA as overall project manager. While these guidelines were not explicitly modified as station planning began in earnest, the very fact of involving non-U.S. entities in that planning implied that other changes in the NASA approach to international cooperation were possible. To a slight but perceptible degree, NASA was recognizing the need to share with others control over shaping potential partnerships.

Space Station Task Force and International Cooperation

In February 1982, NASA Associate Deputy Administrator Philip Culbertson created an informal task force on the space station. This task force was organized around a nucleus of people from the Advanced Programs Office of the Office of Manned Space Flight, in addition to several individuals from elsewhere in NASA. Administrator Beggs on May 20, 1982, formalized the existence of the Space Station Task Force. A major purpose for creating the task force was to make space station planning an agencywide process operating in direct contact with NASA's most senior management, thereby both minimizing the Marshall-Johnson rivalry that had previously pervaded the planning process and involving other NASA cen-

ters in defining the organization's next major project. Named to head the task force was John Hodge, a British-born veteran of the Mercury, Gemini, and Apollo programs who had spent the previous decade working for the Department of Transportation. Hodge had been working with the informal task force members since he had returned to NASA; he had already indicated that he was a strong proponent of international cooperation in the station program.

While the task force had the responsibility for planning the programmatic aspects of the space station, NASA's Office of International Affairs was in charge of developing the policies to guide discussions of international participation in the program. A May 25, 1982, briefing for NASA Headquarters officials captured the state of thinking on international involvement in the space station. Pedersen identified four "key questions to be answered":

- Can such a major project as a space station be undertaken on an international basis and still be effectively managed?

- Don't major international space projects just result in technology leakage abroad?

- Is international involvement consistent with possible U.S. military utilization of the space station?

- What are the *quids pro quo* for foreign contributions to a space station?[24]

This 1982 artist's conception depicts a mature space station configuration, which includes two solar panels to provide power; several modules for command, habitation, and experimental activity; a Shuttle-sized unpressurized rack for the storage of payloads; advanced remote manipulator systems for the assembly of large structures and the servicing/storage of satellites and instruments; and a docking/utility hub that might serve in addition as a "safe haven" in case of emergency. Attached to the station in this picture is a Shuttle orbiter. (NASA photo 82-H-869).

During the early days of task force operations, Hodge created a number of informal working groups. Most addressed technical issues, but two had as their focus more programmatic concerns. One was a "Program Planning Working Group" and was chaired by Robert Freitag. The other was the "International Cooperation Working Group." It also was initially chaired by Freitag, but was soon taken over by Robert Lottmann, although Freitag stayed closely involved. These working groups had as members not only individuals from the task force, but also people from other offices at NASA Headquarters and from various field centers. Throughout the period covered in this study, the International Cooperation Working Group brought together people at the working levels of NASA to discuss international cooperation issues.

Freitag and Lottmann used the working groups as tools for articulating the benefits of cooperation to working-level skeptics throughout the agency. They stressed that the additional financial contribution from potential partners would enhance the scope of the station and that the possibility of international cooperation would increase support for the program overall in the administration and Congress. They also argued that learning to work together on long-term complex projects could form the basis for cooperation on even more ambitious programs in the future.

A series of interactions with potential partners during May and June 1982 had emphasized to John Hodge the high international interest in station involvement. In a July 30 memorandum to Kenneth Pedersen, Hodge noted that "international interest in our space station planning activity is now relatively high. Recent actions by ESA, Canada, and Japan suggest that this interest will be pursued. . . ." Hodge laid out a series of questions that had to be addressed to develop a task force approach to international cooperation, and he asked Pedersen for his ideas on them.[25]

Pedersen's response was a fourteen-page, single-spaced strategy memorandum. In it, he highlighted many of the issues that NASA would have to address in crafting its approach to space station cooperation. Among them were:

1. *When to Involve Other U.S. Government Agencies Interested in International Affairs.* Pedersen noted that "NASA is responsible for making sure that all U.S. Government agencies or portion thereof that have foreign policy responsibilities are kept informed of activities." In carrying out this responsibility, reported Pedersen, NASA was already keeping relevant State Department and Department of Defense offices informed, and had begun to brief the export control community on NASA's planning. He noted that "other agencies such as [the Office of Science and Technology Policy], OMB, DOD, NSC [the National Security Council], and ACDA [the Arms Control and Disarmament Agency] are probably interested in the international aspects as well as the programmatic ones," and he suggested that the Space Station Task Force include those aspects in its briefings to these organizations.

2. *Foreign Reaction to Military Involvement.* Pedersen noted that "this is an important issue, since the interest and debate over the militarization of space is at an all-time high." He thought that it was important for NASA "to be fairly straight forward at all times on the probability and level of DOD involvement expected. . . . We should be working to accommodate both civil and military uses within the basic design of the space station, so that one does not make the other impossible."

3. *Technology Transfer.* Pedersen noted that historically NASA's cooperative programs had been structured carefully to avoid unwarranted technology transfer, particularly by avoiding relationships between U.S. and foreign industry that could lead to such transfers. He thought that "if we carefully choose the cooperative arrangements—for example, we might make sure that they are discrete hardware pieces with minimal interfaces—we can minimize the potential for technology transfer."[26] But Pedersen also noted "growing interest" in the Reagan administration in the topic of technology transfer.[27] He saw "evidence of closer application of existing export guidelines and review of appropriate future steps in stanching the flow of advanced technology," and he recognized the need for NASA to "maintain close and continuing contacts with the export control community."[28]

Pedersen also noted in his memorandum that foreign involvement in the station program would be certain to broaden the project's base of support within both the administration and Congress.

The Allies Are Interested

Even before the formal kickoff of the Space Station Task Force's international activities at a September 13, 1982, "International Orientation

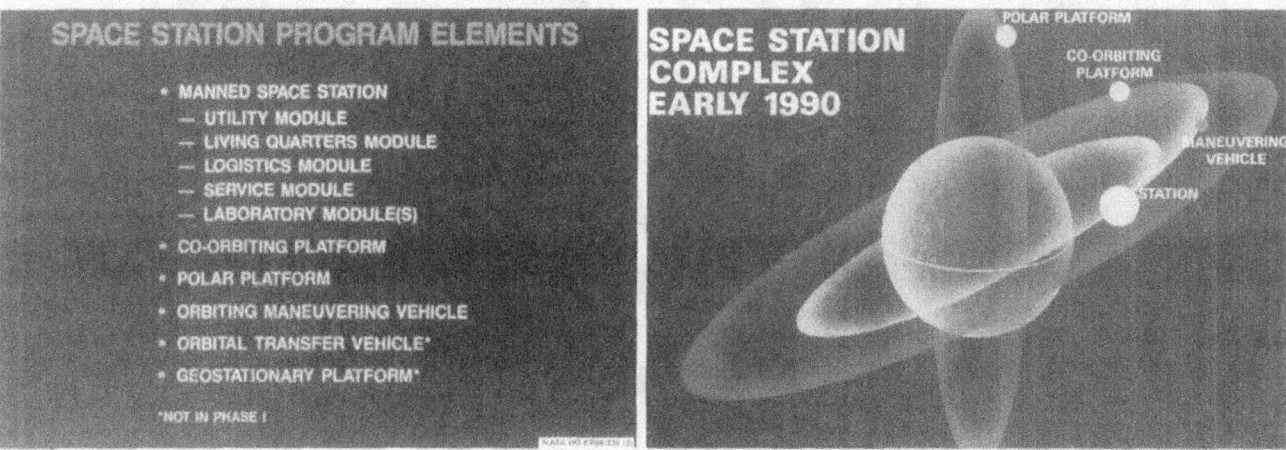

These briefing slides show the space station program elements and the complex as envisioned for 1990. (NASA photo).

Briefing," a substantial amount of non-U.S. planning related to the space station had begun. NASA had adopted a strategy for its own planning efforts of not preparing any particular station design until both the missions such a station would carry out and the capabilities needed to implement those missions were identified.[29] In January, Pedersen, in response to foreign inquiries about how best to proceed, had suggested to potential station partners that they adopt a similar approach.

The response was not long in coming. On May 1, 1982, Japan announced its intent to establish a Space Station Task Force reporting to the top-level Space Activities Commission as its link to the NASA station planning effort; that task force would involve other Japanese organizations as well as the Japanese National Space Development Agency (NASDA) in putting together a plan for Japanese involvement in the station.[30] A mission requirement study was its initial activity.[31] Top-level endorsement of the Japanese effort was provided during a June 1982 meeting in Washington between NASA Administrator Beggs and Minister Nakagawa of Japan's Science and Technology Agency.

On June 17, NASA Administrator Beggs met in Paris with ESA Director General Erik Quistgaard. Among the products of the meeting was an understanding that "ESA will fund, manage, and conduct a first study entitled 'European Utilization Aspects of a U.S. Manned Space Station.' This study will be requirements-oriented."[32]   Canada's National Research Council also initiated a government-industry study of possible Canadian missions that could be carried out aboard a station, but it did not get under way until several months later.

NASA was not totally successful in keeping foreign attention focused on station missions rather than on station hardware. The June NASA-ESA understanding noted that "ESA will fund, manage, and conduct a second study with parallel European contracts, to investigate the European architectural and implementation implications of those requirements identified in the first study,"[33] and Japan and Canada began to study potential hardware contributions as well as mission requirements. In addition, both France and a German-Italian team were studying, independent of ESA, future hardware concepts for a European manned program that could also form the basis for U.S.-European cooperation. This was understandable, noted Pedersen in his August strategy memorandum to John Hodge, because of "'political realities'; they have to justify spending their resources on a space station not only on potential space station utilization but on potential industrial return as well." Nevertheless, urged Pedersen, "we must not let the emphasis on requirements get lost."[34]

Mission Requirements Studies

In August 1982, NASA awarded eight contracts to U.S. aerospace firms to conduct independent but parallel requirements analysis studies. NASA's plan was to combine the results of these eight studies and those being carried out by potential partners to help make the case that a station was justified.

Pedersen addressed the September 13 "International Orientation Briefing" at NASA Headquarters both on NASA's general approach to space station cooperation and its plans for interactions during mission requirements studies. With respect to the former, he noted that the general principles that had guided NASA's international activities in the past, appropriately modified, would apply to the station situation. With respect to the schedule for the mission analysis studies, Pedersen noted that "the mid-term contractors' summaries would be closed reviews. They will be conducted individually with each contractor because NASA does not want the studies to contaminate one another." For this reason, he said, "the reviews will be restricted to NASA personnel." However, immediately following these reviews, NASA would invite "foreign space agency representatives" to hear a NASA summary of the U.S. mid-term results and to present a mid-term status report on their own studies. The final review of the U.S. studies in February and March 1983 would be open, said Pedersen, and he invited the foreign study teams to attend and to present their final results at the same time. Final written reports would be exchanged among all study contractors, U.S. and foreign, in April. In adopting this approach, NASA was hoping to keep not only U.S. but also foreign study teams isolated from each other in the early stages of their efforts. In that way, the thinking went, NASA and the sponsoring non-U.S. space agencies would get the benefit of the independent ideas of all contractors, rather than have the various U.S. and non-U.S. study teams unduly influence one another.[35]

Pedersen also announced that NASA would welcome at any time visits of foreign space agency representatives to NASA Headquarters (but not to NASA field centers) to discuss space station planning. He characterized as "premature" any discussion on potential foreign hardware contributions and modes of cooperation beyond the Phase A planning period.[36]

There were several reasons for NASA deciding to deal only with representatives of foreign governments, and not individuals from non-U.S. industry.[37] For one thing, it was industrial contacts that were perceived as the most likely source of technology transfer. Also, from the start of station planning, NASA wanted to discourage the notion of international teaming during the design and development phases of the program. In the post-Apollo period, U.S. and European industries had teamed to study cooperative possibilities. While such transnational industrial teams, with each firm funded by its own government, ultimately did not emerge in the post-Apollo period, NASA believed that this could have limited its flexibility in structuring the post-Apollo cooperation. While European contractors were selected by ESA in large part to fulfill the requirement that national contributions to ESA be returned to that nation in approximately the same proportion, NASA may not have wanted to select the U.S. partner of a particular European firm as its contractor. The prohibition against foreign visits to NASA centers, even by government representatives, was intended to keep NASA's possible partners from getting involved in intercenter conflicts over station concepts and to

maintain Headquarters control over the international dimensions of the program.

Several aspects of NASA's posture at this time were troubling to potential partners, particularly in Europe. The principles for cooperation spelled out by Pedersen seemed to reflect the same "NASA as managing partner" approach that had been traditional. Europe believed that its accomplishments during the 1970s had earned it a larger voice in future cooperative undertakings. The exclusion of representatives of foreign industry from direct dealings with NASA and the prohibition against visits to field centers to discuss station planning were annoying. NASA's European representative, Richard Barnes, who was sensitive to foreign perceptions, cabled a cautionary message to Pedersen in early September 1982:

> The history of the post-Apollo U.S./European dialogue, as well as more recent experience, suggests that there will be many occasions when a NASA action, taken for legitimate internal programmatic reasons, will be perceived by some Europeans as "evidence" of NASA lack of interest in European involvement. And of course those who want to see European space programs proceed in a direction independent of the U.S. are already looking for such evidence. Let's give them as little opportunity as possible.[38]

However, the undertone of skepticism in Europe regarding NASA's seriousness about desiring significant space station cooperation was not pervasive, nor did it extend to other potential U.S. partners. During the rest of 1982 and early 1983, parallel mission requirements studies went on in the United States, Europe, Canada, and Japan. In addition, the general concept for the space station—known as "distributed architecture"—emerging from studies by NASA and its contractors was particularly congenial to international cooperation. In this concept, the space station would not be a single, large facility, but rather a complex of modules, trusses, and platforms to carry out various space station missions. This made it easier for a foreign partner to contribute a separate space station element that met the criterion of a "clean interface" with other aspects of the station.

A major public symposium was held in mid-1983 to review the status of space station planning. This symposium, organized by the American Institute of Aeronautics and Astronautics for the Space Station Task Force, brought together members of the NASA planning groups, DOD, interested U.S. and foreign industrialists, the press, and representatives from Congress and foreign space agencies. The main purpose of the symposium was to get everyone interested in the station program exposed to the most up-to-date information. In his keynote address, Administrator Beggs noted that the purpose of developing a space station "is, of course, to maintain our leadership." But such leadership would be through cooperation, he suggested, saying that "the space station lends itself uniquely to international cooperation. If we can attract that international cooperation, then other nations will be cooperating with us in the resources that they spend, rather than competing with us."[39]

Summarizing NASA's view of the international dimensions of space station planning to date, Kenneth Pedersen noted that:

[W]e all recognize that the very scope and complexity of the space station process tends to suggest that foreign participation, if it takes place, is going to entail fairly sizeable financial and political commitments on their part. . . . I believe that when and if the time comes that we have the opportunity to proceed "full steam ahead" with the space station, they and their countries are going to be in a better position as a result of this [early involvement in NASA's planning] activity to know what their interests are, to know what their level of participation might be, and in what areas that participation might be most mutually beneficial.[40]

While his assessment of the status of international involvement was positive, Pedersen also added a caution. He noted that "NASA has been aware throughout this space station study that we do not have an approved program. . . . Thus we have not wanted to create unnecessary and unwarranted expectations that would come back to haunt us. . . . We at NASA and the countries that have been working with us have tried to be as open and candid with one another as we could in terms of the state of play, the current situation with respect to the decision-making process, and what we believed a realistic schedule would be."[41]

In mid-1983, this caveat was quite in order. President Reagan in April had directed his top policy-making body for space, the Senior Interagency Group (Space), to prepare over the summer a recommendation on whether he should approve the space station program and include funding for it in the fiscal year 1985 budget.[42] NASA had found few allies within the U.S. government in support of the station; the interagency process was not producing the hoped-for endorsement of the station. The issue of potential international cooperation in a space station was not being addressed at the top levels of the U.S. government; the focus was on the more fundamental policy issue of whether there would be a space station program at all.

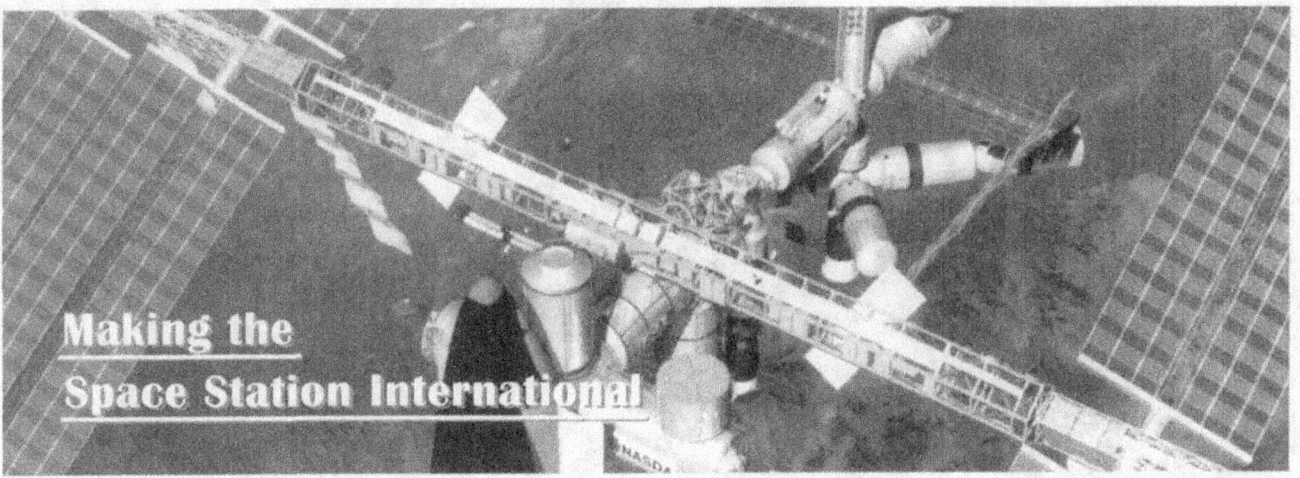

Making the
Space Station International

Resistance to International Involvement Surfaces

By mid-1983, it had become clear to those leading NASA's effort to gain support for the space station among other government agencies that the potential for international involvement was not a strong selling point. From the start, Pedersen and others had recognized that the possibility of technology transfer associated with such involvement would be of concern to the national security community and to administration appointees at DOD. However, they were surprised to discover that the individuals within the Department of State overseeing the foreign policy aspects of science and technology were not enthusiastic about the potential of international cooperation in the space station program to serve broader, foreign policy objectives.

The technology transfer issue first surfaced in terms of 1982 requests by U.S. firms carrying out the space station mission requirements studies to exchange information with their European counterparts. Approval of these requests required the issuance of a Technical Data Exchange Agreement under the provisions of the Munitions Control Act, which was administered by the Department of State. DOD was also closely involved in the approval process.

To lay the basis for the anticipated approval of these requests and to make sure that concerned offices within DOD and the State Department were aware of the overall context of planning for international involvement in the space station program, Robert Freitag briefed officials from DOD's Office of the Under Secretary of Defense for Research and Engineering and the State Department's Office of Oceans, Environment, and International Science and Technology in mid-1982. The reception to the briefings was reported to be "very good," at least in terms of a willingness to listen to what NASA had to say.[43]

However, approval for the data exchanges was not forthcoming. NASA tried to push the process along in

late 1982 by appealing to higher level officials in the State Department and DOD. Talking points prepared for a meeting with DOD noted:

- *There is no need to transfer any sensitive technology at this point. . . . We are not funding sensitive technology such as design details, fabrication or procurement information.*

- *In the RFP [Request for Proposals for the mission requirements studies], even though we did not envision sensitive technology transfer, we wanted to make it clear that the NASA contract award did not constitute approval for any technology transfer. We stated in the RFP that U.S. companies must follow normal export control procedures.*

- *. . . We understand that several U.S. proposals to exchange basic mission needs and general systems information have not yet been approved despite more than four months of review.*

- *Some concerns have been expressed within DOD that consideration of these concerns now is "premature":*
  - *The U.S. has no commitment to a space station program;*
  - *The U.S. has no policy regarding international involvement in a space station program;*
  - *Approval would be a "blank check" for technology transfer.*[44]

NASA believed that the reason for DOD concern was that there had been a change in the individuals controlling the approval process. Apparently, the export licenses had been recommended for approval by the Under Secretary of Defense for Research and Technology, Richard DeLauer, but that recommendation had been rescinded as new

At this meeting of the White House Cabinet Council of Commerce and Trade on December 1, 1983, approval of space station development was the major agenda item. Key personnel in attendance are: Budget Director David Stockman (second from left), Vice President George Bush (fourth from left), Science Advisor George Keyworth (center), President Ronald Reagan (second from right), Secretary of Commerce Malcom Baldridge (third from right), Presidential Advisor Ed Meese (fourth from right), and Gil Rye of the National Security Council (near door). (White House photo C18695-11).

Assistant Secretary of Defense for Policy Richard Perle and his deputy, Stephen Bryen, had been successful in wresting export control responsibility away from DeLauer.[45] Both Perle and Bryen were known as "hard-liners" on technology transfer; having them involved in approving international involvement in the space station did not bode well.

On November 3, 1982, NASA appealed for help to the Under Secretary of State for Security Assistance and Science and Technology, William Schneider, making essentially the same points as had been made to DOD.[46] What the space agency discovered in these and other interactions with the State Department was that neither Schneider nor the Assistant Secretary for Oceans, Environment, and International Science and Technology, James Malone, were supporters of the space station program or of international cooperation in it, and they shared Perle's and Bryen's concerns regarding technology transfer.[47]

Ultimately, DOD's reservations blocked the issuance of export licenses. NASA wrote to the eight mission requirements contractors on December 14, noting that "consideration of these cases within the export control community has become an extended process—the principal concern being that since a space station program has not yet been given a new start, it would be premature to have any *formal* arrangements with foreign industry." Given this situation, NASA suggested that "in the short time remaining until the final report is due in February 1983, we suggest that you restrict your contacts with foreign sources to information which does not require a license."[48]

The recognition that plans for international cooperation could be torpedoed by the opposition of people such as Bryen, Schneider, and Malone was sobering to the advocates of such cooperation, and particularly to Kenneth Pedersen and his top staff person on the space station, Margaret (Peggy) Finarelli. Finarelli had joined NASA in 1981 after tours of duty in the Central Intelligence Agency (CIA), the Arms Control and Disarmament Agency, and the Carter administration's White House Office of Science and Technology Policy. She was thus very sensitive to the concerns of the national security community and their potential for posing an insuperable barrier to NASA's plans. Her sensitivities were viewed as excessive by long-time advocates of cooperation within NASA, and this became a source of some tension between the Space Station Task Force and the Office of International Affairs as the station decision process unfolded in 1983.[49]

The April–December 1983 period was recognized as critical by both advocates of the space station overall and those who wanted the station to be international. Recognizing that strong advocacy of the latter could jeopardize getting approval to go ahead with the station at all, during this period, those heading NASA's interactions with the White House and other agencies chose not to emphasize the international potentials of the program. This approach troubled some of the members of the Space Station Task Force, but it was seen as a tactical necessity by Beggs, Pedersen, and Finarelli.[50]

The Space Station Decision Process and International Cooperation

NASA's first attempt to gain President Reagan's approval for the space station had come in mid-1982. An interagency study of space policy, which began in late 1981 under the leadership of the White House Office of Science and Technology Policy, was nearing completion, and Ronald Reagan was being asked to approve a new statement of national space policy. In addition, Reagan had agreed to attend the landing of the fourth Space Shuttle mission in California on July 4; this would provide an occasion for a presidential statement on space policy. In attempting to convince the White House to announce station approval as part of its new space policy on the occasion of the Shuttle landing, NASA Administrator Beggs wrote Presidential Advisor Edwin Meese in late May. His case for the station stressed its use as both a labora-to-ry and an operations base. He noted the challenge to U.S. space leadership from Soviet, European, and Japanese accomplishments. He argued that a major new project was needed to maintain the human spaceflight development skills of NASA and its industrial partners. As a final argument, Beggs noted that "the space station could also have major foreign policy advantages for the U.S. Both the European Space Agency and the Japanese are interested in participating in its development and would contribute substantial funding if they are given a significant role."[51]

NASA's attempts to gain Reagan's endorsement of the station at this early point were not successful; his advisors thought such a decision was premature. Thus, Reagan's July 4 speech at the Shuttle landing said only that "we must look aggressively to the future by . . . establishing a more permanent presence in space."[52] The station per se was not mentioned.

Perhaps the most important feature of the new National Space Policy announced on July 4 was the transfer of leadership responsibility for developing space policy within the Reagan administration from the Office of Science and Technology Policy to the National Security Council. The policy directive established a Senior Interagency Group (SIG) on Space, chaired by the Assistant to the President for National Security Affairs, "to provide a forum to all Federal agencies for their policy views, to review and advise on proposed changes to national space policy, and to provide for orderly and rapid referral of space policy issues to the President for decisions as necessary."[53] The space station became one of the early items on the SIG (Space) agenda.

Responsible for space policy matters within the National Security Council staff at this time was Gil Rye, an Air Force colonel who had worked on space issues within the Pentagon before being detailed to the White House. While still at the Pentagon, Rye had been the Air Force representative at the NASA space station planning workshop in November 1981, and by 1982, he had become personally convinced that it was in the U.S. national interest to develop a space station. This view was at variance with the Air Force position, which was very skeptical of the value of humans in space and which was centered on making the NASA Shuttle responsive to DOD requirements before any major new NASA initiatives were begun. Having Rye as an ally in the White House proved invaluable to NASA during the 1982–1984 period, both in getting Reagan's approval for the space station and in making international participation a major feature of the station initiative.

Following its inability to gain an early space station endorsement by the White House, NASA decided to wait until 1983 for its next attempt at program approval. A dual strategy was devised. NASA would work through the prescribed SIG (Space) process to attempt to gain the support of other government agencies for the station project, while at the same time NASA's leadership would try to reach the President Reagan and his top advisors directly to convince them of the merits of the undertaking.[54] Meanwhile, the Space Station

Task Force would continue its programmatic liaison activities with potential partners, so that there was a basis for collaborative action should a station program with international involvement be approved.

In support of this strategy, the task force formed a unit called the Concept Development Group. Its task was to integrate the results of field center studies, the eight industry studies of space station requirements, and any input from potential international partners. The chair of the group was Luther Powell of the Marshall Space Flight Center, who had had extensive experience in cooperation with Europe during the Spacelab program. International representatives participated in the activities of the Concept Development Group and were involved in many of the studies of requirements and of systems and subsystems carried out during 1982 and 1983.

In the fall of 1982, SIG (Space) formed a working group on the space station. This group was chaired by NASA's John Hodge, and it consisted of representatives from the State Department, DOD, the Department of Commerce, the CIA, and the Arms Control and Disarmament Agency. Individuals from the Office of Management and Budget and the Office of Science and Technology Policy participated as observers. That group first met in October 1982, and it laid out a schedule that called for a report to SIG (Space) on policy options for the space station no later than November 1983.[35]

It did not take long to discover that there was substantial skepticism among some members of the working group regarding the wisdom of international participation in the station; this skepticism reflected the general attitudes of those at the policy level in DOD and the State Department. The discussion at the group's second meeting on November 22 turned to the issue of State Department approval of exchanges of requirements data between U.S. and European firms carrying out mission analysis studies (as discussed earlier). The State Department representative noted that approval was being delayed even though "there do not appear to be any objections to the merits of the cases," but because "DOD has a concern which it has not yet resolved regarding the broad policy issue of whether there should be international participation in a Space Station."[36]

Most of the SIG (Space) working group's time between October 1982 and April 1983 was spent in developing the specific terms of reference for its study. Once that agreement was reached, Rye decided to elevate the political pressure behind the study

request by having President Reagan, rather than the chair of SIG (Space), sign the terms of reference.

The directive by Reagan that set the guidelines for the formal SIG (Space) study of the space station, signed on April 11, 1983, called out "the foreign policy implications, including arms control implications, of a manned Space Station" as one of five policy issues for examination; international cooperation was not explicitly mentioned.[37] A few days earlier, James Beggs had met with Reagan in a session arranged by Rye. The purpose to alert President Reagan of issues involved in the decision on whether to develop a station. The briefing prepared for Reagan noted that the space station "provides broad opportunity for international cooperation," but this was only one of seven benefits identified as flowing from the station program.[38]

As the study process proceeded in the late spring, it became clear that the Hodge interagency group had become bogged down in technical details and multiple options and was unlikely to produce a policy paper suitable for SIG (Space) consideration. Recognizing this, Gil Rye created a smaller group to develop such a paper.[39] The NASA member of the group was Peggy Finarelli. She continued the approach of downplaying the international aspects of the program; her approach was not totally appreciated by Hodge and others in the Space Station Task Force, who also may have resented her taking over the NASA lead in White House deliberations on the station. While Robert Freitag may have been the most influential of the veteran NASA staffers in pushing for making the space station international and Kenneth Pedersen was the conceptualizer of NASA's approach to station cooperation, Finarelli's tactical efforts over the May 1983–January 1984 period were crucial to creating the domestic basis for the station partnership.[40]

In August, the SIG (Space) process resulted in an options paper for President Reagan on the station program; however, the opponents of the program would not agree to sending the paper forward for presidential decision. Given Reagan's aversion to addressing nonconsensus recommendations, this effectively blocked a presidential decision. In particular, vigorous opposition by Secretary of Defense Caspar Weinberger made it clear that DOD not only would not participate in the station, but also would actively oppose allocating substantial budget resources to a NASA station aimed at civilian uses. The schedule for SIG deliberations had been accelerated to reach a recommendation in time for fiscal year 1985 budget submissions in September, but after a meeting of SIG on August 12, it became evident that a positive recommendation to Reagan to proceed with the station was not likely to emerge from the group.

Given this situation, Rye decided to seek other means of gaining presidential approval. During the September–November 1983 period, NASA's assessment of the prospects for gaining White House permission to move ahead with its highest priority project were very pessimistic, even through the agency had included start-up funds for the project in its fiscal year 1985 budget submission.[61]

Ultimately, NASA's second approach to gaining space station approval—convincing Reagan and his advisors of the merits of the program—bore fruit. Still, international considerations did not play a visible role. President Reagan, through an October 4 National Security Council memorandum, requested NASA to identify its priorities in meeting the goal of space leadership that had been set in the 1982 National Space Policy statement. In his reply, James Beggs said that he was "absolutely convinced that a space station is the next bold step in space. . . . It is an essential piece of our long range plan to reap the full commercial and scientific benefits of space." Nowhere in the response were the benefits of international cooperation mentioned.[62]

Reagan's decision to approve the space station was finally made in early December. Wanting to involve a broader range of agencies in the discussions than just the members of SIG (Space), thereby outflanking station opponents in that body, Rye and another station supporter on the White House staff, Cabinet Secretary Craig Fuller, scheduled a December 1 meeting of the Cabinet Council on Commerce and Trade to discuss the station in Reagan's presence. The model of the space station that NASA prepared for the meeting did not show any foreign contributions to the project.

The meeting went well, and a few days later, NASA learned that President Reagan had given his blessing to the station. However, the issue of whether the space station should be an international effort was not addressed.

Adding the International Element

Although Reagan approved the space station in early December, the question of how that approval would be announced was not decided at that time. Within a few weeks, however, White House political advisors concluded that the station was the kind of long-range initiative that fit into the Reagan's plans for his State of the Union address scheduled for late January. Suggestions on what he should say about the station were solicited by the White House speech writing office, and Rye, Finarelli, and others saw an opportunity to link a presidential invitation for international participation with the announcement of station approval.

NASA was ready to seize that opportunity. Pedersen and Hodge had met as long ago as July 1983 to identify the policy issues that had to be addressed for NASA to proceed with international participation, once presidential approval for the station program was obtained. In a follow-up memorandum, Pedersen had noted the major issues:

1. *What space station "components" are not eligible for cooperation?*

Discussion . . . NASA still needs to decide whether certain elements, while requiring a clean interface, may still be elements which the U.S. should build.

2. *Foreign Involvement in Phase B*

Discussion . . . Should NASA undertake Phase B's on all space station elements, while foreign space agencies fund independent parallel Phase B studies on space station elements in which they have a particular interest? Should NASA entertain Joint Phase B studies?. . . At what point does NASA begin to drive individual countries to particular ele-ments, or should we encourage multiple approaches by all so that natural "fits" fall out?

3. *Guidelines for international participation*

Discussion . . . To what extent do we want to establish de facto minimum contributions (either in terms of funding or in elements)?

4. *Study Agreements, MOUs, and Quids Pro Quo*

Discussion . . . Phase B study agreements would be desirable from the viewpoint of our partners and NASA for several reasons: a) they would provide the framework for information exchange and industry-to-industry relationships; and b) they could strengthen foreign space agencies' position . . . for funding and support.

. . . One major element that must be reviewed now are potential quids pro quo that NASA will want to offer in exchange for hardware contributions. NASA's experience with the [Space Transportation System] program suggests some very good examples that would be appropriate to a space station: NASA commitment to buy additional hardware, preferred access to the space station on a variety of uses, reduced (or

no) costs for utilization, and opportunities for flight of foreign personnel. Of course, formulas for these would have to be worked out so that the benefits match the size of the contributions. . . . In addition, I think NASA should consider international cooperation on the operation of the space station, as NASA and ESA have agreed to do on the Space Telescope, and consider how that should be factored into the equation.

5.　Technology Transfer and DOD Concerns

Discussion . . . Prior to Phase B, NASA needs to develop a set of ground rules for both Headquarters and the Centers on information exchange with our foreign partners. These will not only be useful for reference for NASA employees, but will also demonstrate to the export control community that NASA is aware of the current technology transfer concerns, and doing something about them.[63]

At some point in the fall of 1983, the foreign policy potential of the space station had come to the attention of individuals in the Office of the Under Secretary for Political Affairs and the Bureau of European and Canadian Affairs of the State Department. There was more receptivity to that potential among these individuals than there had been from the science and technology elements of the State Department. As plans for announcing the space station in the 1984 State of the Union address moved forward, Finarelli at NASA and State Department officials Thomas Niles and Arnold Kanter were actively discussing the benefits of station cooperation in the context of broader foreign policy concerns.

These discussions, and the recognition that the issue of international cooperation had to be addressed in some way before approval of the space station program was announced by President Reagan, led to a January 18 meeting convened by the chair of SIG (Space), Special Assistant to the President for National Security Affairs Robert McFarlane, and his deputy, Admiral John Poindexter. Attending the meeting were NASA Administrator James Beggs, Under Secretary of State for Political Affairs Lawrence Eagleberger, Under Secretary of Defense for Policy Fred Ikle, and CIA Deputy Director Robert Gates. This high-level group not only decided to solicit international participation in the space station; they also chose to have the invitation to participate come from President Reagan as he announced his approval of the station program in the State of the Union address seven days later. The top-level group decided that Beggs, acting as the Reagan's

personal emissary, would travel to key foreign capitals to extend the presidential invitation in person. The text of the invitation as it was to appear in the State of the Union address was hurriedly drafted on the evening of January 18 and approved by the meeting participants the next day. There were no interagency meetings or policy papers devoted to the cooperative proposal, nor any formal assessment of the risks associated with international cooperation. This was a decision made by top policy officials, not a ratification of staff proposals. The issue of international participation was not separately raised with President Reagan; his approval came in the form of overall approval of the speech text.[64]

Before he made the State of the Union speech, Reagan sent a personal message to Chancellor Helmut Kohl of the Federal Republic of Germany, President Francois Mitterrand of France, Prime Minister Margaret Thatcher of the United Kingdom, Prime Minister Bettino Craxi of Italy, Prime Minister Yasuhiro Nakasone of Japan, and Prime Minister Pierre Trudeau of Canada:

During my State of the Union address this Wednesday, January 25, I will be announcing the United States' intention to proceed with development of a manned Space Station program. It is my hope that we can work together on this project. To develop this cooperative effort I have asked James M. Beggs, the Administrator of the National Aeronautics and Space Administration (NASA), to act as my personal emissary and meet with senior officials of your government in the near future.[65]

Thus when Ronald Reagan went before Congress on January 25, 1984, and invited other countries to participate in the space station project he had just announced, that presidential invitation came as no surprise to the leaders of those countries that the United States hoped to engage in the station partnership.

Extending the Invitation

The first step in arranging the trip of Administrator Beggs was to develop "terms of reference" to guide him in his meetings. These guidelines were drafted by NASA and circulated for comment by the National Security Council to other agencies that were members of SIG (Space). The staffs of those agencies, which had been bypassed in the rapid process of approving President Reagan's invitation, used this opportunity to make sure that they would be involved in preparing a "report on approaches to international cooperation" for Reagan's approval after the Beggs trip was completed.[66]

The approved terms of reference for the Beggs trip were issued by the President's Special Assistant for National Security Affairs, Robert McFarlane, who was also chairman of SIG (Space). McFarlane wrote Beggs on February 25, saying that "the President would like for you to travel as soon as possible to appropriate foreign capitals as his personal emissary and meet with senior officials to discuss potential international cooperation" in the space station, with the objective being "to agree upon a framework for collaboration on this program which could be announced at the London Summit in June 1984."[67] The idea of including station cooperation as an agenda item on the annual seven-nation economic summit had come from Peggy Finarelli and Thomas Niles and had been embraced by those within the State Department responsible for summit planning.[68]

The terms of reference for the trip specified that in his discussions with foreign officials, Beggs should:

- *Explain NASA's current plans for development of a permanently manned space station, with emphasis on expected capabilities, modular design, anticipated availability, and relationship with the President's overall civil and commercial space program.*

- *Assess the extent of foreign interest in program participation. The assessment should include the level of overall interest, the expected benefits to be achieved, and the foreign resource contributions that might be forthcoming.*

- *During the discussions with foreign officials, the Administrator should avoid making specific commitments regarding international cooperation until other U.S. government agencies have had the opportunity to review the implications.*[69]

President Ronald Reagan announcing the decision to build a space station during the January 25, 1984, "State of the Union Address," while Vice President George Bush and House Speaker Thomas "Tip" O'Neill look on. (NASA photo).

The original plans for the Beggs trip called for the use of commercial airlines. Vice President George Bush, who had offered quiet support for the international initiative all along, suggested to the NASA Administrator that he request the use of one of the Air Force planes available to the White House; Bush indicated that he would support such a request.[70] Accordingly, on February 19, Beggs wrote White House Chief of Staff James Baker requesting the use of a government airplane, arguing that it was "justified and appropriate" because of "the President's direct instruction, the extremely tight timetable, and the importance which space station has assumed here and abroad as a central feature of this Administration's leadership program."[71]

The plane was provided by the White House, and Beggs and an entourage that included Gil Rye from the National Security Council staff, Phil Culbertson, John Hodge, Ken Pedersen, Peggy Finarelli, and Lyn Wigbels from NASA, and Mark Platt and Michael Michalik from the State Department left Washington on March 3. They traveled to London, Bonn, Rome, and Paris and flew directly from Paris to Tokyo, returning to Washington on March 13. After a few days home, the group visited Ottawa. At each stop, Beggs formally reiterated Reagan's invitation to consider participation in the U.S. space station program, and he tried to respond to questions and concerns.

At every stop, Beggs and his group met with space officials and with the highest ranking nonspace officials available, as follows:

- London—with the minister of state for industry and information technology and the science advisor (Prime Minister Thatcher and the foreign secretary were meeting outside of London with French President Mitterrand)

- Bonn—with the minister for research and technology and the under secretary of the foreign ministry (Prime Minister Kohl and the foreign minister were in Washington)

- Rome—with Prime Minister Craxi, the science minister, and the head of the National Research Council.

- Paris—with President Mitterrand, the foreign minister, and the minister of industry and research (Beggs also met with ESA executives and addressed a meeting of that agency's political governing board, the ESA Council.)

- Tokyo—with Prime Minister Nakasone, the foreign minister, and the minister for science and technology (Beggs also spoke to the

Keidanren, the influential federation of Japanese industries.)

- Ottawa—with the minister of state for science and technology, the science advisor, and the president of the National Research Council

One issue in every discussion was the size and cost of the contribution for which NASA was hoping. Beggs had asked Ken Pedersen in January for an estimate of what a reasonable expectation might be. Pedersen's response noted that Europe had contributed approximately 12 percent of the costs of developing the Space Transportation System and that it was "reasonable to expect similar percentage contributions from these countries to Space Station." He noted that the German estimate for a potential space station contribution was $1.5 billion and that Canada was considering a station contribution that "would cost roughly the same" as the $100 million Canada had spent on the Space Shuttle remote manipulator system. Pedersen thought that "it is probably not realistic" to expect Japan's contribution to be half that of Europe, but he noted that the pressurized module that Japan was considering "would cost Japan at least $500 million to develop given their current lack of related [research and development] experience."[72]

Upon his return from Europe and Japan, Beggs wrote to Secretary of State George Shultz on March 16, which summarized his assessment of the trip to date. He told the Shultz that:

The reaction so far to the President's call for international cooperation has been both strongly positive and openly appreciative. It has been positive in the sense that our principal Allies are moving quickly, or have already moved, to take political decisions to participate. And their reactions clearly show appreciation for the major foreign policy benefits that will flow from open and collaborative cooperation on such a bold, visible and imaginative project.[73]

On the basis of the March trips, NASA judged that Italy, Germany, and Japan had in essence already made the political decision, at least in principle, to participate and that France was also likely, after tough bargaining, to be involved. The reception in Great Britain had been the coolest on the trip, and the uncertainty of an upcoming national election made it impossible for Canada to indicate its commitment to cooperation. It seemed as if European cooperation would be organized through ESA, rather than on the basis of bilateral relationships between the United States and specific European countries.[74]

After his round of visits to foreign capitals was over, James Beggs wrote a letter to each country he had visited to summarize his understandings, clarify issues that had been raised, and lay out the next steps. He reiterated the basic U.S. position that:

President Reagan has committed the U.S. to building an $8B fully functional space station to be operational by the early 1990s, but has also set the stage for working together to develop a more expansive international space station with even greater benefits and capabilities for all to use. Thus, we are inviting your Government to take a close look at our plans and concepts and then, based on your long-term interests and goals, share with us your ideas for cooperation that will expand the capabilities of the space station.[75]

In person and in writing, the United States had now extended an invitation for international participation in the space station. Such cooperation had been escalated from a possibility discussed among space agencies to a highly visible initiative of the U.S. president. In the months ahead, the United States would discover whether a framework for accepting that invitation could be developed.

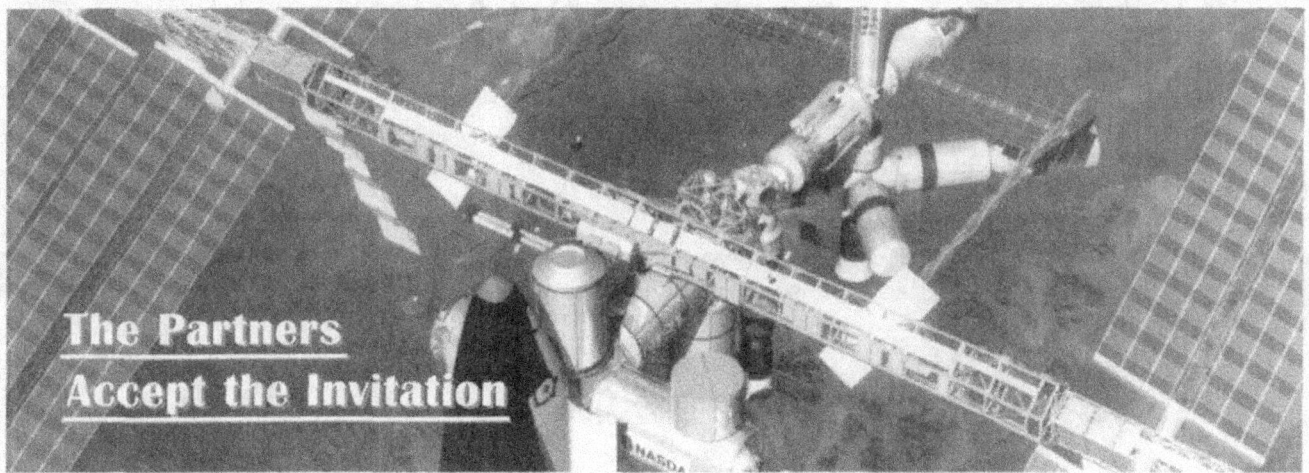

The Partners
Accept the Invitation

Introduction

In the first months of 1984, the hope of the United States was that its invitation to participate in the space station program would be quickly accepted, at least in principle, by political leaders in Europe, Japan, and Canada. It was also hoped that detailed negotiations on the terms and conditions of that participation could then commence, leading to the signing of initial cooperative agreements by the end of 1984. The terms of reference for the trip of James Beggs directed him to seek agreement on "a framework for collaboration" on the space station, "which could be announced at the London summit" in June 1984.[76] Such early agreement was not feasible, however; it took until the first months of 1985 for the political foundation for the station partnership to be established. This section describes the steps that led to European, Japanese, and Canadian acceptance of the U.S. invitation to consider engaging themselves with the space station program.

Early Agreement Sought

The idea of including the station invitation as an agenda item for the London Economic summit arose out of conversations between NASA's Peggy Finarelli and Thomas Niles of the State Department's Bureau of European-Canadian Affairs, after the basic decision to invite international participation in the station had already been taken. Niles remembers that:

Having seen this proposal, my colleagues and I in the State Department who were responsible for the Department's participation in planning for the Summit concluded that this was an appropriate initiative. We based this conclusion on the obvious need for initiatives in connection with the Summit, the fact that the Summit participants were the obvious choices to join with us in the space station, and the reality that

kicking a proposal of the magnitude of the space station up to the Head of State/Government level, through the Summit process, is often the best way to get a decision.[77]

At a January 30 planning meeting for the London Economic Summit, President Reagan approved the notion of asking other summit participants to issue a statement indicating their intent to participate in the space station program. The head of summit preparations in the United States was H. Allen Wallis, Under Secretary of State for Economic Affairs. He and his colleagues in the other six summit countries were known as "Sherpas." The Sherpas met on February 17–19; all seemed open to the idea of having the summit partners declare that they "agree in principle to cooperate in the development of an international space station, demonstrating that free nations will continue to use outer space for peaceful purposes and for the benefit of mankind."[78]

The results of the NASA Administrator's rapid trip, however, suggested that much work would have to be done if any agreement were to be reached in time for the summit. Beggs wrote Wallis that he had come to understand during his trip that: "the Summit declaration is . . . extremely important to NASA's counterpart technical agencies in these other countries. To them it represents the political underpinnings necessary to proceed—analogous to the President's State of the Union guidance to us."[79]

In addition to time needed for each potential partner to develop domestic political support for participation in the space station program, two issues of concern emerged at almost every stop on the Beggs trip. While not insurmountable obstacles to collaboration, they suggested that tough negotiations would be required before final commitment to international participation could be obtained.

One was technology transfer. In his follow-up letter to those he had met on his trip, Beggs recognized that "technology transfer has been an increasing concern on all our parts in the past few years, and we will need to work together to make sure we are protecting our mutual technology bases in this partnership."[80] The other issue of general concern was the extent of U.S. military involvement in the space station. Here, the U.S. position had been carefully crafted to reflect both anticipated foreign sensitivities and to be acceptable within the U.S. government. Beggs told potential partners that:

> The U.S. space station program is a civil program which will be funded entirely out of NASA's budget, with no national security funds used. . . . The space station that the President directed NASA to build is a civil space station. Of course, like the shuttle, the space station will be available to users. If there are any national security users, like national and international users, they will be able to use the facility. As provided in the Outer Space Treaty, however, all activity on the space station will be limited to peaceful, nonaggressive functions.[81]

Beggs also reported that "our principal allies are moving quickly, or have already moved, to take political decisions to participate."[82] This may have overstated the situation somewhat. On one hand, having the invitation to participate come from the U.S. President and be extended to other heads of government had changed the stakes. The preceding two years of discussions at the technical level, and the biases toward collaboration that had emerged from those discussions, were transformed into an issue high on the policy agenda. No ally wanted to be in a position, without compelling reasons, to refuse President Reagan's public invitation. On the other hand, all three potential partners—Japan, Canada, and Europe—were in the midst of their own internal debates over the future direction of their space efforts. Accepting the U.S. invitation, even in principle, implied that a significant share of their space budgets over the coming decade would have to be channeled into a partnership with the United States. Beggs had made it clear that the U.S. desire was for significant contributions to the station, roughly equivalent to 10 to 20 percent of the partners' overall space budgets for the next decade. Whatever their leanings toward accepting Ronald Reagan's invitation, in few of the potential partners had there yet been enough discussion to make their leaders willing to make a firm political commitment to collaboration of that character and scope.[83]

As a followup to the Beggs trip and in preparation for the summit, Gil Rye, Peggy Finarelli, and Robert Freitag made an April trip to Europe, meeting with both space agency officials and summit Sherpas. Their discussions reinforced the sense that some in Europe would be cautious about making a commitment to cooperation at the summit. They also found that the smaller member states of ESA, which were not part of the summit process, were concerned about a summit declaration that could commit them to additional contributions to ESA. There was limited enthusiasm for the station proposal in some of these states, both because their industries did not see the prospect for significant business in the undertaking and because finance ministries, almost always opposed to increasing space budgets, had more influence than space advocates in smaller ESA member countries.

The potential for international participation in the U.S. space station was a "talking point" on President Reagan's agenda for his private meetings with each of the other six leaders at the London Economic Summit, which took place June 7–9; the issue was not discussed during the formal plenary sessions of the summit leaders. However, as they emerged from one of those meetings, the seven leaders encountered a large model of the station that (unlike the model that NASA had brought to the White House the preceding December) included representations of potential foreign contributions; this was a carefully staged opportunity for President Reagan to discuss his invitation to participate. NASA's Langley Research Center had prepared the detailed station model, which the U.S. summit delegation (including Gil Rye and Peggy Finarelli) had carried to London; twenty to thirty minutes of lively discussion and a "photo opportunity" ensued as the summit leaders gathered around the model.

The summit communiqué was cautious in its language, saying (in its final substantive paragraph) that:

> We believe that manned space stations are the kind of programme that provides a stimulus for technological development leading to strengthened economies and improved quality of life. Such stations are being launched in the framework of national or international programmes. In that context each of our countries will consider carefully the generous and thoughtful invitation received from the President of the United States to participate in the development of such a station by the United States. We welcome the intention of the United States to report at the next Summit on international participation in their programme.[84]

The London Economic Summit of June 7–9, 1984, during which the space station was a major topic of discussion. Left to right: President Reagan (United States), Prime Minister Margaret Thatcher (United Kingdom), Foreign Minister Graf von Lambsdorf (Germany), and Prime Minister Yasuhiro Nakasone (Japan). (NASA photo).

While this statement was less of an endorsement than had been proposed to the summit Sherpas in February, the noncommittal language of the communiqué accurately reflected the state of affairs in June 1984.[85] Even so, it was an endorsement of the station concept and thanked Ronald Reagan for his invitation. The inclusion of station cooperation on the agenda for the 1985 summit was particularly significant. It was intended to encourage speedy decision-making in Europe, Japan, and Canada, because any delays or breakdowns in discussions over acceptance of President Reagan's invitation would have to be reported back to the summit leaders at their next get-together. Although more time would be needed to find ways in which the U.S. invitation and the separate space goals and ambitions of Europe, Japan, and Canada could be combined in ways acceptable to all partners, there was now a deadline to provide a focus for deliberations around the world.

Even creating an initial agreement to work together in seeking such a melding of interests, capabilities, and programs would require separate negotiations between the United States and each of its potential partners. Europe, Japan, and Canada had

been engaging in informal discussions with the United States regarding possible space station cooperation since January 1982. However, as they made their own space plans, they certainly had not been able to count on the station gaining the early and unambiguous approval of the Reagan administration that was communicated by including approval of the program in Reagan's State of the Union message. President Reagan's approval of the station and his invitation to participate changed the context in a major way. Europe and its major countries active in space—France, West Germany, Great Britain, and Italy—as well as other potential U.S. partners were making their own plans and decisions based on their own interests, and the role of large-scale collaboration with the United States had to be evaluated in terms of those interests. As one close observer of the European space scene remarked (and his remarks were in many ways applicable to Japan and Canada as well):

The dilemma which faces countries of Europe as America's space station program begins to get underway concerns chiefly priorities, both national and European. To maintain Europe's existing space programs and take on a new space station activity would require a major increase in space-related expenditures and thus a reappraisal of national priorities.

To some, it might appear that the U.S. would be called upon to provide guarantees and accept dependence in excess of what Europe's share of the common burden will be worth. But the imbalance is the other way: Any substantial European involvement in a U.S.-led space station program would absorb so much of the space budget that Europe would forfeit the ability to create a similar but independent capability.[86]

The invitation by Reagan to participate in the space station program had been a true leadership initiative; it was now up to the potential partners as to whether they chose to follow the U.S. lead.

Europe Charts Its Future in Space[87]

Much had changed in Europe since the post-Apollo agreement to develop Spacelab as part of the U.S. Space Transportation System. European commitment to the German- and Italian-led Spacelab program had been part of a "package deal" among countries interested in space. Other elements of that deal were multilateral funding of a French-led program to develop an independent launcher for Europe, Ariane, and a British-led program to develop a maritime communications satellite. In addition, eleven European nations had agreed to create a single organization to manage programs in science, applications, and infrastructure development. By the time the U.S. invitation for space station participation was extended, Spacelab had had a successful first flight aboard the Space Shuttle. Ariane was in service and successfully launching both government and commercial payloads, and the maritime satellite was in operation, serving as the initial basis for the INMARSAT organization. ESA had developed into an effective means of combining the resources of member states to support programs that not one of them was able to carry out on a unilateral basis; ESA programs combined with national efforts had led to the emergence of a vigorous space industrial base in Europe.

ESA Planning Includes a Cooperative Option

Considerations of future programs were very much on the European agenda in the early 1980s, as the efforts begun on the basis of the 1973 package deal approached completion and European industry, national space agencies, and ESA assessed ways of building on past achievements. The possibility of European involvement in a U.S. space station pro-

gram was part of these considerations from the start. Indeed, as long ago as 1976, being aware of early U.S. space station studies, the ESA Council (the organization's "Board of Directors" composed of representatives from its member states) had resolved that ESA should "examine the questions connected with a possible participation by Europe in the Space Station programme."[88]

In February 1982, NASA Administrator James Beggs and ESA Director General Erik Quistgaard discussed potential NASA-ESA cooperation on the station program; each directed their head of advanced planning (Ivan Bekey for NASA and Jacques Collet for ESA) to work together as station planning gained momentum.[89] Based on this guidance, a detailed plan for NASA-ESA coordination and joint activity regarding station planning was quickly developed; Europe was thus given the opportunity to be involved in the station program almost from its inception.[90] By June 1982, ESA and NASA had agreed on an approach in which ESA would carry out two sets of space station–related studies. One, to be called "European Utilization Aspects of a U.S. Manned Space Station," would be conducted in parallel to U.S. mission requirements studies; other ESA studies would investigate the architectural and implementation implications of European requirements—that is, what hardware made sense for Europe to contribute to a station program. In September 1982, ESA awarded the contract for the utilization study to the German aerospace research establishment, DFVLR, and initiated four contracts with European industry regarding potential hardware contributions.

These initial steps in European consideration of station participation were taken in anticipation of ESA member-state approval of a "Space Transportation Systems Long-Term Preparatory Programme" (STSLTPP) that would provide the overall context for charting Europe's future plans in the area of launch and in-orbit systems. The STSLTPP had been approved in principle by the ESA Council in June 1982.[91] It was intended to provide member states "the elements necessary for making decisions on the selection of a long-term policy and on the start of new programmes" to follow Ariane and Spacelab. Among the options to be analyzed by the STSLTPP were "investigation and preparation of the necessary decision elements on: maintaining in Europe an independent launch capability, developing a European in-orbit infrastructure, and pursuing transatlantic cooperation through participation in the future United States space station programme." One of the three "themes" to be investigated was how to "provide Europe with a capability of carrying out orbital operations (including return to Earth) by means of in-orbit infrastructures developed independently or by cooperation with NASA in the future U.S. space station activities."[92]

While staff members of ESA may have welcomed the possibility of continued cooperation with the United States, their attitude was not universally shared in Europe. It proved difficult to get member-state commitment to the STSLTPP, in large part because of skepticism in some countries regarding the wisdom of continuing intimate cooperation with the United States. NASA European Representative Richard Barnes reported in December 1982 that the ESA Council had "again deferred, this time for a month, the deadline for member states adherence to the . . . STSLTPP which includes funding for Ariane 5 and Space Station studies. So far only Sweden, Belgium, Denmark, and Germany have formally signed up, with Germany the only strong supporter of Space Station studies."[93]

In the weeks following the December ESA Council meeting, advocates of at least examining cooperation were able to gather the support needed for carrying out the STSLTPP. France agreed to support the study program on December 22, Italy on January 6, 1983, and the United Kingdom on January 14. With the four major ESA members signed on, study efforts were able to go forward during 1983 and 1984. Commenting on the adoption of the program, the leading French newspaper, *Le Figaro*, noted:

The old continent is preparing its space activity for the next century: we will undoubtedly have then our space-men, orbital infrastructure and maybe, also a mini-shuttle to fly on our own. At least that's what ESA—who is initiating an important engineering program in this regard and has already signed the first industrial study contract—thinks. Hopefully, we will know between now and 1985. At the same time we will know who will influence this long-term policy: Germany who favors complete cooperation with NASA, or France, more favorable to independent solutions.

At the moment, two philosophies are possible. On the one hand, the one of German industrialists that consider that Europe should work in full cooperation with the U.S. . . . From there, however, opinions diverge: The French, in fact, would like to keep a certain "independence" as far as manned flights are concerned and thus conduct studies in such a manner as to preserve the means to equip Europe with a complete [Space Transportation System] to embark men. The problem is, one can imagine, that it would be very expensive. . . .

Thus, the main task is to convince our European partners of the value of those expenses. However, in order to succeed, France will first have to resolve its own contradictions: Some of us still believe that the space exploitation will be a reality by the end of the century without any human presence, which is counter to the future outlook on both the American and Soviet sides. As long as such opinions carry weight in France, it will certainly be difficult to claim to be able to influence ESA's decisions.[94]

Not all early thinking about space station cooperation was carried out within the ESA framework. Another focus for considering potential European contributions to the U.S. space station emerged from studies carried out by Germany and Italy. Interest within the two countries in using Spacelab hardware as the basis for future programs dated back to the late 1970s. Advocates of continued cooperation with the United States, particularly within Germany, sought an approach that would preserve the option of cooperation, either through ESA or outside of it. In 1983, the German firm, MBB/ERNO, and the Italian firm, Aeritalia, under the respective supervision of DFVLR and the Italian CNR (the national research agency in charge of the Italian space plan), began intense studies of the use, either in conjunction with the U.S. space station or as an independent European-controlled orbital complex, of an orbital infrastructure consisting of Spacelab-derived pressurized modules, unmanned platforms, support modules, and service vehicles. The name given to this orbital complex was Columbus; the program was "viewed by some countries as a German/Italian effort to secure the lead role in Europe's space station development."[95]

The French space agency, *Centre Nationale d'Études Spatiales* (CNES), and the French aerospace industry also were studying future space efforts in the early 1980s. One focus of attention was a new high-thrust rocket engine, designated HM60, designed to use liquid hydrogen and liquid oxygen as fuels; such an engine would be used to develop a new generation of the Ariane launcher, designated Ariane 5. In other studies, attention was given to an automated or human-tended space station concept called Solaris and to a small winged spaceplane called Hermes. As NASA began space station studies in 1982, CNES set up its own examination of station mission requirements. The goals of this study were to allow CNES "to determine independently its interest in cooperating with NASA on a future space station; but also to determine whether it is in their best interest to cooperate through ESA or directly with NASA."[96] Although the French government had traditionally been

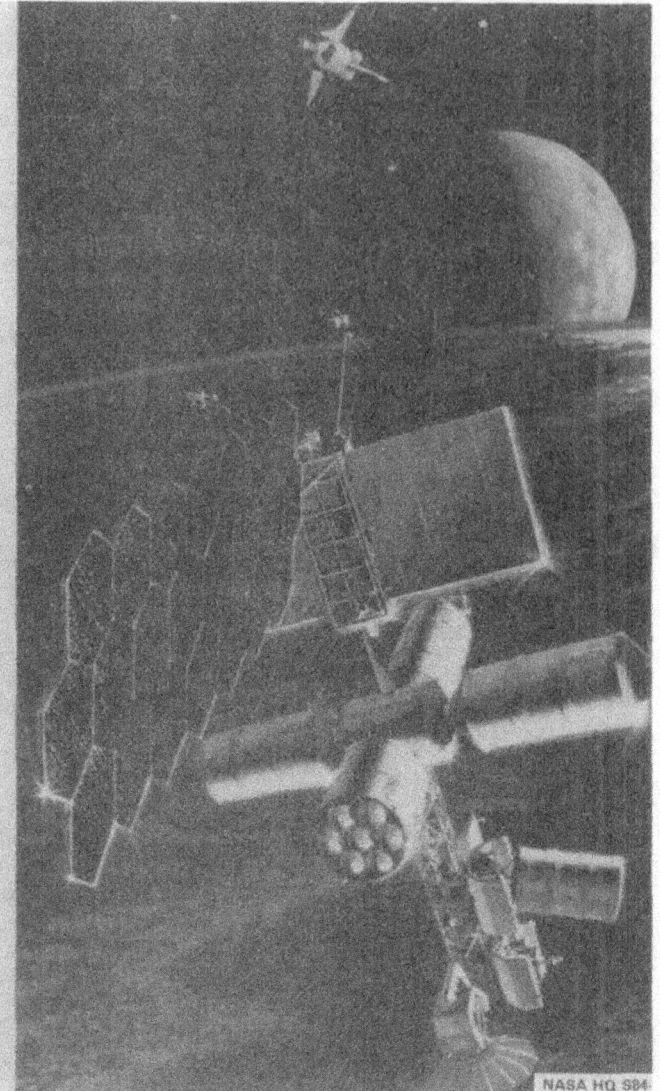

**SPACE STATION:
WHY NOW?**

- **DEVELOPMENT PROGRAMS OF SPACE SYSTEMS ARE LENGTHY**

- **SHUTTLE BECOMING OPERATIONAL**

- **SOVIET THREAT TO U.S. CIVIL LEADERSHIP IN SPACE IS REAL**

- **A PROSPEROUS ECONOMY REQUIRES INVESTMENT IN NEW TECHNOLOGIES**

A late 1984 graphic on the rationale for building the space station. (NASA photo HQ S84-2032A(3)).

skeptical of the importance of human spaceflight activities, this attitude shifted 180 degrees following President Francois Mitterrand's decision to accept a Soviet invitation to fly a Frenchman aboard the Soviet *Salyut* space station. That flight took place in June 1982. France from this time on increasingly argued that independent European capabilities in all areas of space activities, including human access to orbit, were essential; the term "autonomy" was used to describe this ability to act without dependence on others. The appropriate balance between European autonomy and intimate engagement with the United States became a major issue in the 1983–1985 debate over European space policy.

In 1984, Germany and Italy proposed to their ESA partners that Columbus be considered as an optional program[97] to be carried out within the ESA framework; France did the same for the HM60

engine. These proposals were approved in principle by the ESA Council on June 28, 1984. ESA was authorized to attempt to gain member-state financial commitments for preparatory studies, prior to a final decision to proceed, on the development of the large cryogenic HM60 engine and on a "space station related programme based on the proposal by the German and Italian delegations . . . this programme will be defined with a view to ensure progressively the European autonomy in the field of manned space station compatible with the future European launching systems." The Columbus preparatory programme would also include "consideration of the invitation received from the President of the U.S."[98]

The ESA staff spent the remainder of 1984 incorporating the Columbus program and plans for a new European launcher into an overall long-range European space plan. (They were also working closely with the NASA space station planners to stay abreast of U.S. activity, now that the space station had received President Reagan's approval.) Other inputs into this plan came from the results of the STSLTPP and from the planning activities of other offices within ESA concerned with science and applications programs. That long-range plan was ready for initial consideration by ESA member countries in November. The introduction to the plan noted the need to find the right balance between:

1. Science and applications—between cultural and economic rewards

2. Payloads and launchers/in-orbit infrastructure—between ends and means

3. Launcher development and manned space flight—between major technological avenues, that of propulsion and that of human-in-orbit

4. Manned space systems and automated space systems—between humans and robots

5. ESA program and national program—between centralized and decentralized activities

6. Purely European program and cooperative ones, in particular with the United States—between achieving space autonomy and undertaking large-scale programs and their operation

The ESA executive alerted member states that "the present scope of the overall ESA programme will have to be enlarged, making it necessary to increase the funding at an average rate of 12 percent a year over 5 years."[99]

The plan recommended that Europe develop a new launcher, Ariane 5, based on the HM60 engine, to become operational by the end of 1995. It noted that the U.S. space station was a "major step in space capability which Europe cannot afford to ignore" and recommended "until about 1995, to improve through cooperation with NASA the existing European manned flight operations." To this end, the plan proposed approval of the Columbus program, "involving cooperation with the U.S. in the development, operation and utilization of an international space station, subject to negotiation with NASA of satisfactory terms and conditions for such cooperation." In conclusion, the ESA plan suggested that "the most urgent task ahead is for Member States to reach a broad consensus, within the ESA forum, on a well-balanced

and ambitious programme for the next ten years, derived from a shared vision of Europe's future in space."[100]

Developing Political Support for Station Cooperation

In fact, the elements of such a consensus had been emerging in Europe during 1984. The June ESA Council decision to approve the HM60 and Columbus preparatory programs had foreshadowed a new "package deal" to guide Europe's next decade in space. While studies sponsored by ESA and national space agencies defined possible hardware elements of the next generation of European space capabilities and of potential European contributions to the U.S. space station, political-level discussions among the leading European countries—particularly France, Germany, the United Kingdom, and Italy—were leading to agreement on how those elements could be combined in an acceptable fashion. A key to these discussions, in addition to agreement on the hardware elements to be included, was developing an understanding on how various ESA member states would distribute among themselves the costs, and the proportional industrial involvement, in the various elements of the ESA plan.

The major difference of view that had to be resolved in these discussions was between the long-standing French preference for an emphasis on improved launch systems and for an approach that stressed European autonomy and the German and Italian preference for both continued development of human spaceflight capability and continued close cooperation with the United States in that development. Another consideration was the British preference for ESA to undertake applications programs that produced tangible benefits, rather than research or exploration-oriented activities. For most of the smaller ESA member states, a primary concern was a program with enough diversity and breadth to allow meaningful opportunities for their scientific and industrial participation. These differing preferences had been accommodated in the 1973 package deal that had guided European space activities for a decade; during 1984, the political support for a similar combination grew. The need to respond to the U.S. invitation to participate in its space station program certainly accelerated the process of agreement and shaped its content, but the desire for a new European commitment to its future in space was an equally influential stimulus.

A meeting of the ESA Council, at which each member state would be represented by its cabinet minister responsible for space activities, was scheduled for the end of January 1985. This was the first European space meeting at the ministerial level since the 1973 gathering that had created ESA and approved the Ariane and Spacelab programs. The purpose of the January 1985 meeting was to consider the long-range plan proposed by ESA. In the course of putting together that plan, there had been close consultation among the ESA Director General[101] and senior members of the ESA executive staff and policy-level officials within the governments of ESA member states. The European aerospace industrial organization, Eurospace, had put together a proposed long-term European space program that reflected the views of its industry members; it was in essence the same as the proposed ESA long-range plan, suggesting that European industry was ready to lend its support to the ESA proposals.[102] The major unresolved question as 1984 drew to a close was whether those proposals would receive the political and financial support needed to move ahead. Ultimately, it was up to the individual ESA member states—and particularly France, Germany, and the United Kingdom (Italy had already answered in the positive)—to decide whether they wanted to increase their financial and political commitment to space, and to ESA, to the levels required to carry out the program that ESA was proposing.

The British Position. At the time that President Reagan first invited international participation in the station program, the United Kingdom was perhaps the most skeptical of the major ESA member states regarding both a significant increase in the ESA budget and significant European engagement in the U.S. space station effort. These would require additional funds at a time when the Thatcher government was giving overriding priority to cutting the U.K. budget. In addition, early discussions of potential European contributions to the U.S. space station had not clearly identified any element or activity of particular interest to Britain.

This latter issue was resolved during 1984. The concept of the German-Italian Columbus program included one or more automated platforms to carry scientific and applications instruments. The leading U.K. space firm, British Aerospace, became interested in having the lead role in supplying these platforms—particularly an Earth-observing platform in polar orbit to complement a similar polar platform that was part of the "distributed architecture" of the U.S. station concept.[103] An important feature of European space planning is the ability to reach informal agreements on which a country's firms would act as prime and secondary contractors for various ESA programs in advance of their actual approval. Germany and

Italy agreed to allocate to Britain and to British Aerospace the lead role in the platform aspects of the Columbus program, and this provided the incentive the British government needed to go along with the proposed ESA long-range plan and European participation in the U.S. space station program. Even so, some degree of skepticism about the appropriate priority of space activities overall and of ESA programs in particular lingered among some in the Thatcher government and the British bureaucracy, although Thatcher herself was visibly enthusiastic about the station program, once she had been briefed on it in preparation for the London Economic Summit. In fact, at the summit meeting, it was Thatcher who had taken the lead in the discussion as the seven leaders gathered around the space station model.[104]

To build a broader base of support for the space station program within Britain, the U.K. Department of Trade and Industry, under whose auspices the space program operated, organized an October 4, 1984, meeting on the station program. The new Minister of Trade and Industry, Geoffrey Pattie, told the meeting that the government "had no preconceptions" and thus was very interested in the opinion of attendees on whether Great Britain should support station cooperation within ESA; the tone of Pattie's remarks to the symposium, however, were quite positive toward station participation. A summary of the meeting noted general agreement that "the Space Station is a logical development" and "surprising unanimity that we should go ahead via ESA."[105] One NASA official visiting the United Kingdom in the fall of 1984 reported that top British space officials "appeared optimistic about Cabinet approval for a major British contribution to the ESA Space Station program."[106] When the U.K. cabinet did meet in late 1984, it decided to provide those funds and to make the accompanying commitment to cooperation with the United States on the station program.

The French Position. Italian support for station cooperation and for the ESA long-range plan incorporating it had never been in question. What was uncertain as the January 1985 ESA ministerial meeting grew closer was whether France and Germany could find an approach to Europe's future in space that reflected the interests of both countries. Earlier, the outlines of a French-German compromise that would enable agreement on future ESA programs had been evident in the June 1984 ESA Council approval of preparatory programs for the HM60 cryogenic engine and for the Columbus program. The proposed ESA long-term plan was based on these central features.

France complicated the situation in late 1984 by requesting "Europeanization" of its Hermes space-plane, arguing that just completed internal French studies had demonstrated the feasibility of the concept and that the goal of European autonomy was not achievable unless Europe had its own means of access to space for human crews. This last-minute push for Hermes was a surprise to most and was not well received by space advocates in other major European countries, particularly Germany and Great Britain, because it implied a higher cost for the overall space "package" that they were already having some difficulty selling to their finance ministries. In France, by contrast, strong support for space came from Francois Mitterrand, who had early on in his presidency accepted the Gaullist notion of space as an arena in which to demonstrate French *grandeur*. In addition, the French Minister of Research, Hubert Curien, had been head of CNES before being appointed as Minister, and he was actively pushing his counterparts in other countries for approval of the Hermes concept.

France had been advocating Hermes since the middle of 1983; the Mitterrand government had decided to make the space plane a key element of a plan for French space preeminence in Europe. One justification for Hermes was that it could give Europe independent crew access to the space station, so that Europe did not have to be totally dependent on the Space Shuttle. This was part of a more general French strategy of offering Europe an alternative to dependence on a close alliance with the United States as a key to its space future, should discussions on station cooperation falter.

When France pushed its partners during 1984 to include Hermes development in the package to be considered by the January 1985 ministerial meeting, both the United Kingdom and Germany resisted, believing that there had been inadequate study of the concept to justify a commitment to its development and being less committed to the politically driven concept of European autonomy in space than was France.[107] France continued to advocate Hermes right up to the time of the ESA ministerial meeting. During January 1985, Fredric d'Allest, CNES Director General, made a tour of European capitals in an attempt to increase support for the concept. On January 29, just two days before the meeting convened, a column by d'Allest titled "A Space Policy for Europe" appeared in the influential French paper *Le Monde*. In it, d'Allest argued:

Participation—with conditions yet to be negotiated—in the American space station through the Columbus project would allow Europe to benefit earlier from the use of the

space station by conducting experiments that she would find useful and at low cost.

However, the sour experience of Spacelab cooperation, as well as the U.S. policy to limit technology and technical information transfer to the bare essentials to insure the compatibility of European and U.S. elements, indicate the limitation of such a cooperation. That is why a European policy in this field cannot count heavily on cooperation with the U.S.

Because of the major stakes involved, France has the same determination as she did 10 years ago in Brussels when she convinced her European partners to build Ariane. France proposes a fundamental new step forward in European space programs by deciding, right now, to acquire its autonomy in manned space flight and the progressive establishment of a European Space Station.[108]

The German Position. While support for space station cooperation with the United States had always been strong among German space officials and in the German aerospace industry, during the 1982–1983 period, political support for the undertaking was not yet assured. However, events in late 1983 changed this situation.

One of those events was the visit to Washington of Dr. Heinz Riesenhuber, German Minister for Research and Technology. Prior to Riesenhuber's meeting with James Beggs, Ken Pedersen told the NASA Administrator that the minister "was reportedly very favorably impressed with the amount of public interest in space which was generated by the visit of the [Shuttle test vehicle] Enterprise to Bonn in May 1983." Pedersen noted that "one purpose of discussions with Riesenhuber is to promote station activities, especially international cooperative activities."[109]

Based on his discussions in Washington, Riesenhuber became an enthusiastic advocate of station cooperation; during the following years, he became an essential U.S. ally in securing European participation in the station program. Returning to Bonn, he wrote James Beggs on October 27 that:

While I am aware that there is no approved program, I am interested in coordinating with you as early as possible, even prior to the final decision, the possibilities of a European participation in a space station in the now ongoing preparatory phase. I would be quite willing to take the initiative as to the point that the Federal Republic of Germany, based on her

*coordinating responsibility in the Spacelab
cooperation, will provide the necessary political
and programmatic prerequisites for a European
participation in the space station.*[110]

The support of Riesenhuber, and indeed of
German Chancellor Helmut Kohl, for continued
space cooperation with the United States was reinforced by the first successful Spacelab flight in late
November 1983. President Reagan and Kohl
engaged in a three-way live conversation with the
Spacelab crew, which included the first ESA astronaut, German citizen Ulf Merbold. After the mission, Kohl wrote Reagan that the mission should be
seen "as a symbol of our joint future."[111]

The importance placed by Riesenhuber on
assuring political support within Europe for station
cooperation was a critical factor in the European
decision-making process during 1984. Late in
1984, Riesenhuber and French Minister of
Research Hubert Curien came to agreement on
German support for Ariane 5 and French support
for Columbus, if only their respective governments
would approve the budgets required; this "space
summit" was a critical step in clearing the path for
the Rome ESA ministerial meeting.

However, even with Riesenhuber's strong support
and the long-term bias toward cooperation with the
United States in space, German support for the ESA
long-range plan was not assured as the Rome ministerial meeting drew near. A major sticking point
was budget. A German commitment to the large
programs proposed by ESA—Columbus and Ariane
5—implied either an increase in the German space
budget overall or a reallocation of the resources of
Riesenhuber's Research and Technology Ministry.
German space scientists (echoing their U.S. colleagues) were skeptical of the scientific value of the
space station and strongly opposed to a reduction in
the space science budget as a means of financing it.
German Finance Minister Gerhard Stoltenberg, on
the other hand, resisted increasing the Research and
Technology Ministry's budget to allow for the additional funding required. Another factor that came
into play was the preference of Foreign Minister
Genscher for closer Franco-German ties rather than
continued emphasis on the transatlantic German-
U.S. alliance.[112]

This potential deadlock within the government of
the strongest European supporter of station cooperation was worrisome to the United States. On
December 13, President Reagan wrote Chancellor
Kohl, reiterating U.S. hopes that Germany would
agree to participate in the station project.[113]

The controversy was settled in early January
when Riesenhuber and Stoltenberg agreed to a compromise. Enough additional funds would be provided to the Research and Technology Ministry to cover
half of the costs of German participation in the
Columbus and Ariane 5 programs; the Ministry
would reprogram some of its existing budget to support the rest of the cost of those programs. The U.S.
Embassy in Bonn reported that the German cabinet
would "make a final decision on space station participation at a January 16 meeting. This will be little
more than a formality, since the Chancellor is
known to be in favor of the Space Station and has
only been waiting for his ministers to agree on a
financing plan."[114] The cabinet did meet on January
16 and agreed to German participation. A press
release announcing the cabinet decision noted that
"our cooperation in space research is an important
step on the way toward European integration and
continuous improvement of transatlantic friendship." Prerequisites for successful cooperation,
noted the statement, included:

- *assurance of an appropriate relation between
give and take,*

- *guarantees for access and necessary services,
such as transport with the space shuttle,
support and data transmission under
nondiscriminatory conditions,*

- *guarantee of unlimited scientific and
commercial utilization of results gained,
unrestricted technology transfer for the
development of ESA's own contribution and
for the commercial utilization of
instrumentation and results, and options for
the utilization of European launcher
capabilities.*[115]

The reference to the use of European launchers
did not imply German support for Hermes. In a
press conference following the cabinet meeting,
Riesenhuber said that it was "premature" to decide
on a commitment to Hermes, but that Germany
had not ruled out participating in the program
sometime in the future.[116]

*ESA Ministers Approve Long-Range Plan, Station
Cooperation*

Despite this rejection of French aspirations, at
least for the time being, by the beginning of January
1985, the U.K., French, and Italian governments had
indicated their intent to approve the ESA-proposed
plan, which at the time included no mention of
Hermes. With the German approval of its participation in the Ariane 5 and Columbus projects, the last

obstacle to approval of the ESA long-range plan, and to political agreement to the principle of European participation in the space station program, had been cleared. The ESA Council met in Paris on January 23 in preparation for the ministerial-level meeting the following week, and it discovered that the agency's member states were in agreement on all essential decisions to be taken at Rome.

Meeting in Rome at the ministerial level on January 30–31, 1985, the ESA Council accepted the proposals of the ESA executive for a long-range European space plan and agreed to the initial two-year commitment of funds required to carry out that plan. In so doing, it approved a statement of objectives for the ESA program that included, among other goals, the intent:

- to strengthen European space transportation capacity, meeting foreseeable user requirements within as well as outside Europe, and remaining competitive with space transportation systems that exist or are planned elsewhere;

- to prepare autonomous European facilities for the support of man in space, for the transport of equipment and crews and for making use of low Earth orbit; and

- to enhance international cooperation and in particular aim at a partnership with the United States through a significant participation in an international space station.[117]

Although the French proposal to include Hermes in the approved ESA program at the same level of commitment as Ariane 5 and Columbus was rebuffed, the ministers left the door to future Europeanization of Hermes wide open, taking note with interest of the French decision to undertake the spaceplane program and the proposal by France to associate her European partners interested in this program. The ministers invited France and associated partners to keep the agency informed of progress of these studies with the view of including this program, as soon as feasible, in the optional program of the agency.[118]

With these decisions, Europe committed itself to an ambitious future space program of its own and accepted, subject to the negotiation of acceptable terms and conditions, the U.S. invitation to participate in what ESA insisted on describing as an "international space station," rather than a U.S. station with foreign participation. (The differences in these two characterizations were more than semantic, because the degree of non-U.S. participation and the consequent share in the content

and control of station development and operations were unsettled issues as far as Europe and other participants were concerned, while within the United States a decision that America must have the dominant station role had already been made.) With respect to what acceptable terms and conditions of a space station partnership might be, there had been, at least since 1983, clear indications of the European position. Preliminary discussions on a NASA-ESA agreement had been under way for several months in anticipation of a positive outcome at the ministerial meeting.

The European ministers at Rome went on record as to the objectives that had to be met if the space station partnership were to be viable in European eyes. Those objectives were stated in the form of an ESA resolution that was not made public until it had been delivered to both President Reagan and NASA Administrator Beggs. The resolution noted, with respect to the U.S. invitation, that the ESA Council:

Accepts that offer—with a view to contributing and strengthening a genuine partnership in the space field with the United States of America . . . subject to the achievement of the following fundamental objectives:
- to seek an appropriate European participation by the Agency in the space station programme;
- to give Europe responsibility for the design, development, exploitation and evolution of one or several identifiable elements of the space station together with the responsibility for their management with the aim of increasing the overall capability of the space station;
- to ensure that Europe may have access to and use, on a nondiscriminatory basis, all elements of the space station system on terms that are as favourable as those granted to the most-favoured users and on a reciprocal basis;
- to reach a satisfactory agreement on the share of the operation costs of the station;
- to reach a satisfactory agreement on the level and conditions for the appropriate transfer of technologies;
- to ensure that supplies and services provided by the United States industry and NASA for European requirements are offset by European supplies and services;
- to ensure maximum legal security and an identical level of the commitments entered into by the European Governments and the United States Government;
- to guarantee the availability of American transportation and communication facilities required for the programme and the possibility of using the European facilities as they become available for the programme.[119]

This statement of objectives identified almost all of the issues that would have to be resolved in what turned out to be three more years of detailed and difficult negotiations between the United States and Europe—and also with Japan and Canada—in creating the final framework of agreements for the original station partnership.

Because U.S.-European space cooperation in the space station would be based on a longer, more intense, and sometimes difficult history than with other prospective partners,[120] and because the anticipated European contribution to the partnership would be approximately twice (in financial terms) that expected from Japan and more than four times that expected from Canada, European acceptance of the U.S. invitation was an important achievement for those within the United States advocating the station partnership. Without European involvement, the partnership they had in mind would have been much different in character.

Even after the Rome meeting, there was lingering opposition to station cooperation in Europe. France made it clear that it was ready to take the lead in a program leading to European autonomy, should space station negotiations run into major obstacles. Smaller ESA member states, who in general did not see industrial return proportionate to the costs to them of station participation, remained skeptical. However, the political strength of an invitation from the U.S. President kept this opposition muted in character. Only if the terms laid down by the United States for participation were unacceptable was it likely that Europe would refuse to be the primary partner of the United States in the space station program.

Japan Determined Not to "Miss the Boat"[121]

Once the United States had formally invited Japan to participate in the space station program, there was little doubt that invitation would be accepted. During the 1969–1970 period, the United States had asked Japan to become involved in the planned post-Apollo program of manned spaceflight. At that time, Japan was just getting started on a large-scale space program, even though it had been carrying out small scientific space activities throughout the 1960s. It took Japan some time to form the internal consensus required to respond positively to the U.S. invitation; by the time its response came, the conditions for post-Apollo cooperation had so changed that the opportunity Japan had decided to pursue was no longer available.[122] So Japan was excluded from any opportunity to work with the United States in the human flight area during the 1970s. When the

chance to become involved in the space station appeared, according to one informed observer, "Japan—politicians included—does not want to miss the boat. The Japanese space community wants to participate."[123]

President Reagan's invitation to participate in the U.S. space station program came at a difficult time for Japan, however. The country's space budget, after rising rapidly during the 1970s, had shown a slight decrease in 1983 over 1982, and no meaningful growth was planned for 1984.[124] Even so, a revised space development policy for Japan had just been proposed, and an implication of that policy was an increased Japanese commitment to space over the longer term. A central guideline of that policy was "establishment of autonomy." The policy proposal noted that:

In the space technology field, Japan has been relying on advanced foreign nations in its large portion because of her later starting, and the activities have been under the great influence of such advanced nations.

Japan should, however, establish its own technological capacity for its space development in the future so that its broad and diversified space development activities can be performed in a steady manner.

At the same time, Japan should possess advanced capability in order to implement space development activities properly at its discretion.[125]

The "advanced nation" referred to in the policy proposal was the United States, and the "technological capability" important to Japanese autonomy included both an indigenous launcher and an indigenous satellite bus incorporating advanced technologies. The United States had helped Japan develop launch vehicles and satellites during the 1970s by licensing U.S. industry to sell various technologies to its Asian ally, but those licenses carried limits on how advanced the technology thereby transferred could be. The objective was to license only that technology less advanced than the current U.S. "state of the art." In addition, Japan could not launch non-Japanese payloads using the boosters employing licensed U.S. technology without explicit U.S. permission to do so.[126]

Japan recognized, as had Europe a decade earlier, that its independent access to space was a precondition for any degree of autonomy and, in early 1984, was in the final stages of deciding to develop a new launch vehicle, to be called the H-II, based totally on Japanese-developed technology.

The H-II program was aimed at a first launch in 1991, and it was estimated to cost almost $1 billion to develop.[127] To accept President Reagan's invitation meant that Japan would have to commit itself to even more increases in its space budget, because the kind of contribution the United States was asking would be as expensive, if not more expensive, than the projected cost of the H-II program. While space advocates in Japan were enthusiastic about the possibility, the space program did not have a high priority outside of the science and technology community, and it was uncertain whether the government would be willing to make the financial commitment required to carry out both the H-II and space station programs.

The Japanese space community had been considering its response to a possible U.S. invitation since NASA had raised that possibility in early 1982. In August of that year, it had established an Ad Hoc Committee on Space Station Programs, reporting to the blue-ribbon Space Activities Commission that advised the prime minister on space policy. In typical Japanese style when considering a new area of activity, the membership of this committee included representatives from various government ministries and their national research institutes, Japanese industry, and academic institutions with potential interest in the space station program. This step was also taken to indicate that Japan considered the space station to be a project of government-wide interest, not just the concern of one Japanese agency.

Japanese space activities were carried out by two separate organizations. One, the Institute of Space and Astronautical Sciences (ISAS), was totally devoted to space science; it had evolved from a University of Tokyo group and was still quite academic in style. ISAS received its relatively modest funding from the Ministry of Education, and it cherished its independence from the rest of Japanese space efforts. The bulk of Japanese space work was carried out under the management of the National Space Development Agency (NASDA), which was a public corporation operating under the policy guidance of the Science and Technology Agency (STA), although it received funding from other government ministries and public corporations as well as from STA. In mid-1982, the Space Activities Commission formally designated NASDA as the lead agency in Japan for space station planning.[128]

On an informal basis, Japan had been examining possibilities for involvement in the space station program from the start of 1982. On July 16 of that year, the government established a space station task force;

that group managed Japan's mission requirements studies that were carried out in parallel with similar NASA studies and other station-related investigations.[129] By October 1982, NASDA was able to join with several other organizations to sponsor a space station symposium in Tokyo; almost 400 attendees heard 92 papers presented.[130]

Japanese industry was quick to get involved. By September 1982, the Mitsubishi Group had briefed the government on its concepts for participation, which included an "Experiment Module" consisting of "a manned pressurized module and an unpressurized pallet." According to Mitsubishi, Japanese participation would:

- *Establish a Japanese base for future space activities by participating in the U.S. Manned Space Station . . . Program*

- *Enlarge the field and scale of Japanese space utilization activities*

- *Invest and participate in the rapidly progressing advanced space technology*

- *Contribute to the international society in a worldwide cooperative space development era*

- *Activate and promote Japanese manned space activities.*[131]

(The Mitsubishi presentation so impressed NASA Administrator James Beggs that he sent a copy to Secretary of State George Shultz as an example of the benefits of international cooperation in the space station, which would "provide an opportunity to attract foreign research and development funds into a program which is perhaps uniquely mutually beneficial. . . ."[132]) Other space industries in Japan also studied the concept of an attached experimental module; also under investigation was a Japanese contribution in the form of a free-flying, unmanned experimental platform.

In March 1983, the Space Activities Commission and its Ad Hoc Committee on Space Station Programs met with a NASA delegation led by Associate Deputy Administrator Philip Culbertson (to whom the NASA Space Station Task Force reported). The NASA team also met with people from STA, NASDA, and ISAS. The main purpose of these meetings was to provide Japanese officials concerned with the space station an in-person, top-level view of NASA's space station planning activities and to indicate how Japanese activities fit into those plans. The NASA delegation stressed at every opportunity that the station program had not yet been fully defined, much less approved by the White House and Congress. With regard to specific Japanese hardware contributions to the station, the NASA representatives noted that the Japanese "still had a lot of work ahead to prove to themselves—and us—that they should undertake developments of this scope."[133]

On June 15, 1983, the Ad Hoc Committee on Space Station Programs of the Space Activities Commission issued an interim report that identified materials processing, life science, and advanced technology development as the uses most likely to benefit from the existence of a space station. The committee thought that a module attached to the space station was the best site for work in these areas, and it concluded that "a very large space system can be built" and that "the space station is the first step of the enlargement of the living space of human being[s]."[134] Throughout the year, Japanese interest in station participation continued to increase. For example, in October the influential paper, *Nihon Keizai Shimbun*, reported that:

> *Nissan Motor Co., Ltd. will participate in the U.S. space station program in collaboration with Hitachi, Ltd. and Fuji Heavy Industries, Ltd. . . . The Mitsubishi Group of firms have already announced their policy of actively participating in the program. . . .*
>
> *The Japanese government plans to participate in the space station project from the beginning, that is, even as the project is in the development stages.*[135]

By the time that James Beggs formally extended President Reagan's invitation to participate to Japanese Prime Minister Yasuhiro Nakasone in March 1984, acceptance of that invitation was a foregone conclusion, if acceptable terms for that participation could be developed and if the Japanese Finance Ministry and the Diet (the Japanese legislature) could be convinced to provide the additional funds required to support the cooperative undertaking. In the revision of Japan's space development plan unveiled in late February 1984, which gave the go-ahead to the H-II rocket, the Space Activities Commission had also indicated Japanese intent to participate in the station program. At the time of the Beggs visit, the Prime Minister Nakasone and the science and technology minister made it clear that Japan would participate in a meaningful way, but that government statements in support for the program would remain low-key until the process of developing consensus within Japan had taken place.

As part of the process of consensus-building within Japan, during 1984, five industrial groups within

Japan—Mitsui, Fuyo, Sumitomo, Mitsubishi, and Nisho-Iwai—formed teams to study station utilization and hardware development opportunities. Within the government, the influential Ministry of International Trade and Industry (MITI) became involved in space station–related activities by creating its own study committee on space environment utiliza-tion. Both of these developments caused conflict. *Nihon Keizai Shimbun* reported that there was disagreement within the private sector over which industrial group should have the lead in Japan's involvement in the space station and that the MITI move into space was viewed by STA as an incursion into its area of jurisdiction.[136]

In reaction to this situation, on November 19, the powerful Keidanren (a federation of Japanese industries) formed a fifty-four-member Ad Hoc Committee for Promotion of Japanese Participation in the Space Station Program. The purposes of this group were "(1) to unify the space station use research groups . . .; (2) to coordinate the views of the private sector; and (3) to coordinate information with the . . . STA and the . . . MITI." This move toward creating consensus was seen as essential if STA was to get the budget allocation required to participate in Phase B definition studies for the station.[137]

Indeed, it was the approval of this budget, rather than any formal announcement, that would signal Japan's acceptance of the U.S. invitation. By mid-1984, STA had decided that Japan's contribution to the space station should be an Experiment Module, and studies of several other possible hardware elements were halted. In December, NASA confirmed to STA that such a module would be an acceptable Japanese contribution.[138] After negotiations with the Ministry of Finance that had begun in August, on December 28, STA "with great pleasure" notified NASA that the budget for Phase B station activities had been approved within the government and would be sent to the Diet in January.[139]

Members of the Keidanren space station study committee visited the United States in February 1985 to hear for themselves U.S. responses to a variety of questions that had been raised about the station program. They were apparently satisfied with what they heard. Upon his return to Japan, the leader of the team, Tadahiro Sekimoto, President of NEC (Nippon Electric Company), wrote Administrator Beggs, telling him that "as the Space Station Program is an international project under your initiative, I hope you would go ahead with it by way of cooperation . . . between the two countries. We will, of course, do our best on our side to promote the Space Station Program."[140]

The Japanese Diet approved the funds for Japanese Phase B station activities in April 1985; with that approval, Japan became the last of the three potential U.S. partners to make the political commitment to attempt to find an acceptable framework for cooperation. (Canada had announced its intention to participate on March 18.) In its final report, issued in the spring of 1985 and reflecting the thinking that led to the Japanese acceptance of the U.S. invitation, the Ad Hoc Committee on Space Station Programs of the Space Activities Commission identified the benefits Japan saw in participating in the station:

1. *Acquisition of highly advanced technology: It is expected that the space station will utilize highly advanced technologies in broad areas and, therefore, through the program Japan will acquire extremely advanced technologies such as manned support technology, assembly technology for a large structure in space, etc., and also will encourage development of various advanced technology areas in robotics, computers and communications. This effort is expected to contribute to the advancement of technical standards not only in space but in many other technical fields.*

2. *Promotion of the next generation science and technology coupled with expansion of space activities scope. . . .*

3. *Contribution to international cooperation: Japan's space development policy attaches importance to harmonizing Japanese national space development activities with international space activities. . . . Japan's participation and co-operation in the [space station] program will be quite effective in maintaining and further promoting the friendship between the United States and Japan, coupled with contributing to the elevation of Japan's own technology, by working with the space development activities of the free world.*

4. *Encouragement of practical use of space environment: The realization of the space shuttle regular flights in the United States has strongly pushed forward experiments in the space environment for the production of materials and pharmaceutical products using the microgravity of space. . . . The expansion of commercial activities to space is now a target of various overseas countries as well as the United States and, therefore, this aspect has significance.*[141]

As with Europe, then, Japan's decision on its participation in the U.S. space station was part of a larger set of decisions on future Japanese interests in space overall. Also similar to Europe, Japan recognized that it could not both accept the U.S. offer and satisfy its other space objectives without increasing its financial commitment to space. Finally, as with Europe, Japan saw as its ultimate goal autonomy defined in terms of independence of action in critical areas of space activity. But unlike Europe, there were no influential skeptics within Japan regarding the wisdom of accepting the U.S. invitation, although the Japanese space science community expressed little interest in becoming involved with the station program. The intense consultations and analyses within and between government and industry from 1982 to 1984 had produced a consensus in support of intimate collaboration with the United States in exploring the potentials of human spaceflight.

Canada Sets Its Space Priorities[142]

Although Canada had been actively involved in space since the 1960s and had provided the Remote Manipulator System (also called the Canadarm) as an integral element of the Space Shuttle, the country in the early 1980s had no central space agency. Also, since the 1960s, it had renounced any ambitions related to independent access to space through a Canadian launch vehicle. Thus one of the considerations influencing European and Japanese evaluation of the U.S. space station invitation—the desire to achieve substantial autonomy—was not relevant to the Canadian situation. Canada, to be active in space, had to cooperate; the issues were with whom and on what projects.

Planning for Canada's space activities was the responsibility of an Interdepartmental Committee on Space, chaired by the Ministry of State for Science and Technology. It included as members those ministries that were potential developers and users of space capabilities. Many space-related research activities were funded and managed through the National Research Council of Canada, a quasi-independent government corporation. The National Research Council had been Canada's link to NASA for the Remote Manipulator System project. The Ministry of Communications and the Ministry of Energy, Mines, and Resources also had substantial space involvement.

When the United States invited Canada in 1982 to begin to think about participation in the space station program, other projects seemed to the members of the Interdepartmental Committee on Space to have higher priority as Canada shaped its space plans for the second half of the 1980s and beyond. In particular, two large (for Canada) projects directly related to Canadian needs—a Mobile Communication Satellite for links among Canada's widely dispersed population and a Radarsat for Earth observations through cloud cover—were top-priority projects. In its initial evaluations, the Interdepartmental Committee on Space gave potential Canadian involvement in the U.S. space station the lowest priority among these projects.

The factors that changed this ranking were primarily political in character. The intense public interest in the visit of the Space Shuttle test vehicle *Enterprise* to Canada in June 1983 demonstrated to Canadian politicians the symbolic importance of involvement in human spaceflight. At the same time, the Canadian government announced that it would accept the U.S. invitation to have a Canadian astronaut fly aboard the Space Shuttle.[143] While within the Interdepartmental Committee on Space the Ministry of Energy, Mines, and Resources continued to advocate the Radarsat program and the Ministry of Communications continued its support of the Mobile Satellite Program, from mid-1983, the Ministry of State for Science and Technology had the political advantage through its link to human spaceflight.

In contrast to the U.S.-European experience with Spacelab cooperation, U.S.-Canadian cooperation on the Remote Manipulator System had been a very satisfactory experience on both sides. Karl Doetsch of the National Research Council, who had managed the Remote Manipulator System project for Canada and was one of those supporting station cooperation, remarked in mid-1983 that "there's a good feeling that comes to the fore immediately, which says that the space station is great, we want to be a part of it. . . . However . . . one needs a little more than that. One needs to find good reasons for it."[144]

To this end, the National Research Council of Canada sponsored station utilization studies, as had other potential station partners. Two studies were conducted, one by Spar Aerospace Ltd. and the other by a consulting group, Philip A. Lapp Associates. The studies concluded that "Canada could benefit scientifically, technologically, economically and socially through participation in the development of the Space Station."[145] Particularly attractive to many Canadian users was the existence of a polar orbiting Earth observation platform as part of the station program, because data from remote sensing was important to many Canadian applications. Also, Canada saw the station, with its requirements for in-orbit assembly and operations, as an opportunity to build on the Remote Manipulator System program and

to develop further Canadian industrial capabilities in automation and robotics.[146] Summarizing the position of station advocates within Canada, Doetsch said:

We also think that the space station as a development and as a technological stimulant has strong justification in its own right. This is coupled with the needs of the users, but it mustn't be forgotten.

The rate of return on investment is important, but the strategic benefits are also important.[147]

Canadian-U.S. coordination at the technical level continued during 1983 and 1984, but the political decision on whether to accept President Reagan's invitation to participate had to be put on hold. At the time that James Beggs and his entourage visited Ottawa in March 1984, Liberal Prime Minister Pierre Trudeau had announced that he would leave office, and Minister of State for Science and Technology Donald Johnston told the U.S. delegation that a Canadian response to the Reagan's invitation could not be given until the elections were over, because it was the next government that would have to make the financial commitment to back up an acceptance of the invitation.[148]

A Progressive Conservative government headed by Prime Minister Brian Mulroney was elected in September 1984; that government was philosophically more attuned to the Reagan administration than had been its liberal predecessor and thus was more likely to be positive toward accepting the U.S. offer as a means of strengthening U.S.-Canadian relations. Canadian astronaut Marc Garneau flew as a payload specialist aboard an October 1984 Shuttle mission, further reinforcing the Canadian desire to be involved in future manned activities.

In a paper prepared for a December 1984 NASA international workshop on the station program, Karl Doetsch summarized the "principal issues governing Canadian participation":

- *Importance to Canada of the use of the space station.*

- *Importance to Canada of the privileged access to the infrastructure which will accrue to participating nations.*

- *Desirability of maintaining and enhancing Canada's existing area of industrial space leadership.*

- *Importance of spinoff to Canadian industry in the chosen areas of development.*

- *Extent of the return on investment to be derived from participation.*

- *Desirability of cooperating with other major nations in a major international venture which will have a profound effect on man's ability to exploit the space environment.*[149]

These were clearly very different considerations than had stimulated Europe and Japan to consider participating in the station project.

The technical and the political arguments in support of accepting the U.S. invitation proved ultimately persuasive, but only after lengthy and intense discussions within the Interdepartmental Committee on Space. As one individual closely involved in both the internal Canadian discussions and those between the United States and Canada commented:

It was a judgment call. It was the result of endless discussions. . . . There was certainly a fair amount of unhappiness that it [the station participation] was going to run away with all the funds. . . . It was a matter of visibility. If the space station was going to there Canada had to be part of it. This was the line of argument that was used. And the potential benefit to industry—that helped push it through.[150]

One factor influencing at least the timing of the Canadian decision on whether to accept the U.S. invitation was the first summit meeting between President Reagan and Prime Minister Mulroney, scheduled for Quebec City on March 17–18, 1985. In late January, the top Canadian space policy official, W.M. "Mac" Evans, was optimistic that there would be a positive decision on the part of the Canadian cabinet by that time.[151] The U.S. Department of State welcomed this news; it noted that it and other agencies were involved in an "exercise pointing towards achievements that can be realized before or during the March 17–18 summit in Quebec. Canadian cooperation on the [manned space station] would be such an achievement. . . . But we have not discussed the possibility with the Government of Canada." The State Department noted that "an announcement at the summit need not necessarily be lengthy."[152]

The Canadian cabinet's Committee on Economic and Regional Development did approve a recommendation for Canadian participation in the space station on March 5; full cabinet approval followed quickly thereafter. The approval came in the context of an endorsement of "Canada's Interim Space Plan, 1985–1986," a document that had been prepared by the Interdepartmental Committee on Space. This plan noted that:

The government has decided to accept the invitation of the United States to participate in the Space Station Program. . . .

Space Station will be the predominant civilian space initiative of the remainder of the century and will alter dramatically many of the established ways of operating in space. Canadian participation would permit us to maintain and improve our competitiveness in a number of leading-edge space technologies. All of our international partners have decided to participate which will offer us further opportunities to develop new business relationships and cooperative programs with the world's major space nations.[153]

When Ronald Reagan and Brian Mulroney completed their "Shamrock Summit" on March 18, their joint communiqué announced that "the Prime Minister informed the President that Canada has accepted the U.S. invitation to participate in the space station project."[154]

The overall Interim Space Plan was announced on March 20. Its interim nature was very evident; it provided funding for the three potential major space projects (mobile satellite, Radarsat, and the space station) only for the 1985–1986 period. It noted that "final decisions" on these projects would be taken at the end of 1985, when a long-term strategic plan for Canadian space efforts would be issued. Future funding for the three major space projects would be determined in accordance with "strategic thrusts" set out in the long-term plan.[155]

The development of the Canadian long-range space plan and the assignment of priorities to the three competing projects proved very contentious, although Canadian involvement in the space station would eventually gain top priority. But that was in the future. With the March 18 announcement of the Canadian decision to accept the U.S. invitation, all three partners—Europe, Japan, and Canada—had made the initial political commitment required. Now it was up to representatives of the prospective partners to determine whether a framework for cooperation acceptable to all could be developed.

Conclusion

Creating that framework eventually required three rounds of international negotiations. One created a set of three memoranda of understanding (MOU) that would govern interactions between NASA and its prospective partners during the preliminary design phase (Phase B) of the station program, while more a permanent framework for those interactions was created. This round of negotiations was completed on June 3, 1985, when NASA and ESA signed their Phase B MOU at the Paris Air Show. Japan and Canada had agreed to similar MOUs earlier in the spring of 1985. These negotiations were not particularly contentious; NASA and its partners agreed to defer to the next negotiating round attempts to resolve the kind of difficult issues that had been identified in the January 31, 1985, ESA Resolution 2 on space station cooperation discussed above.

The second round of space station negotiations *was* highly contentious, and on several occasions its successful outcome was in doubt. However, on September 29, 1988, the United States, a number of European countries, and Canada signed an intergovernmental agreement on station cooperation (Japan signed the agreement later), and NASA signed more specific MOUs with its counterpart space agencies in Europe, Japan, and Canada for cooperation during the detailed design, development, and operation and utilization phases of the space station program.

Five years later, in December 1993, the original space station partners decided after the end of the Cold War and the collapse of the Soviet Union to invite the Russian Federation to join the station partnership. There followed another four years of difficult discussions to revise the station intergovernmental agreement and MOUs to accommodate a major new partner; the new agreements were finally signed by all partners except Japan on January 29, 1998. (The approval processes within Japan again were not completed in time for Japan to sign the agreements at that time.)

This account does not cover the space station negotiations between 1985 and 1998. It is perhaps too soon to trace the various compromises that were made by all parties to the discussions in order to reach understandings to which all could agree, and it of course is too early to make a definitive judgment on the success of the partnership.[156]

When the United States and its closest allies began, in the early 1980s, to consider an ambitious international partnership to design, develop, operate, and utilize a permanent space laboratory—

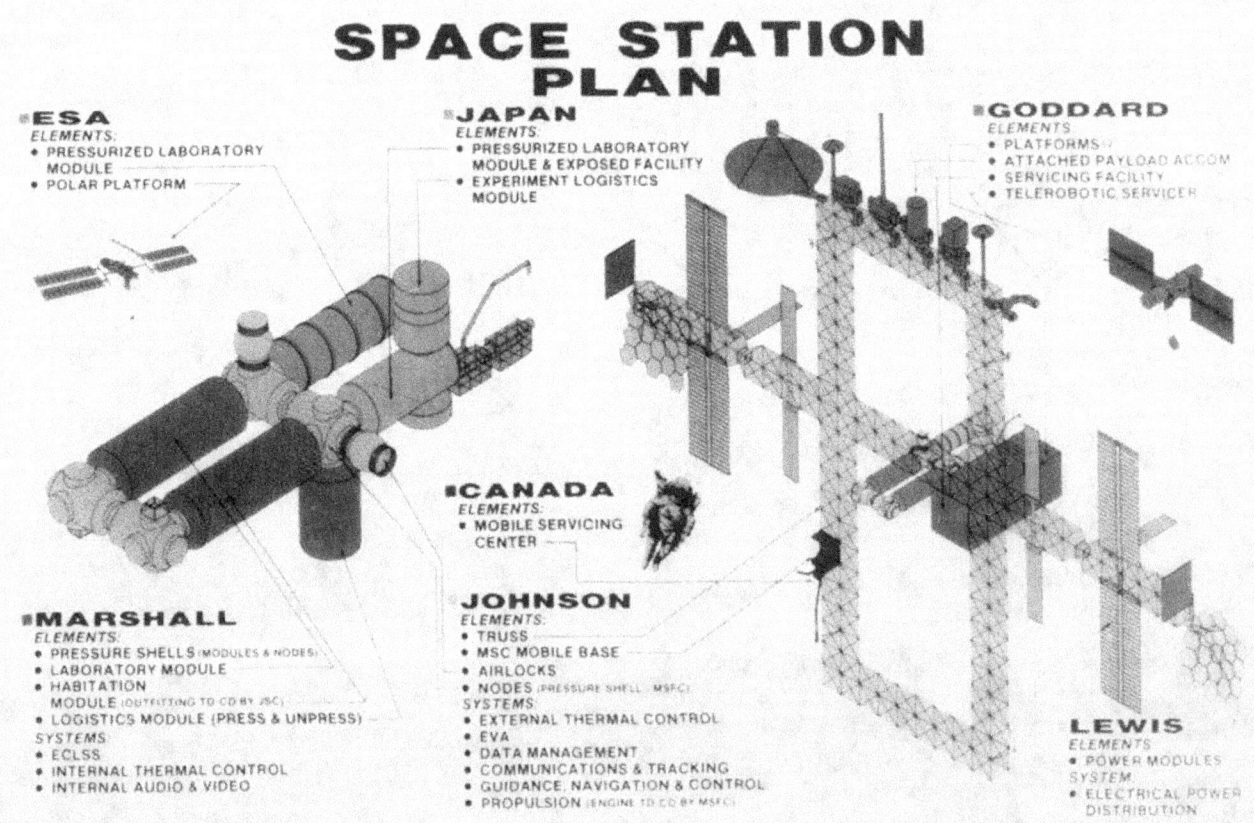

The Space Station Plan, as proposed in 1986, at the time of the initial agreements for international participation. (NASA photo 86-H-324).

what has become known as the International Space Station—they could not possibly have anticipated the twists and turns in the road to making that partnership a reality. When President Ronald Reagan announced his approval of the space station program in January 1984, he directed NASA to complete the undertaking within a decade. It is likely to be *two* decades after Reagan's announcement before all elements of the International Space Station are in place and ready for use. One can only hope that the results of the partnership that began with both high anticipation and mixed feelings, in what was a different era in space development, justifies all the time and effort to make it a success.

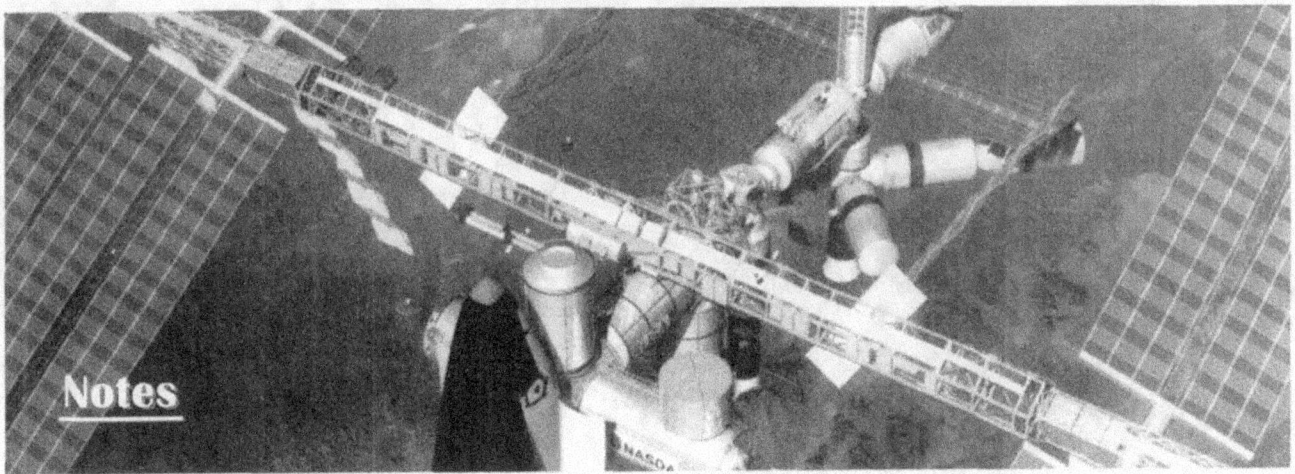

Notes

1. White House, "Address by the President on the State of the Union," January 25, 1984.

2. See Chapter One of John M. Logsdon, gen. ed., with Dwayne A. Day and Roger D. Launius, *Exploring the Unknown: Selected Documents in the History of the U.S. Civil Space Program, Volume II: External Relationships* (Washington, DC: NASA SP-4407, 1997), for a discussion of this change in strategy and for documents related to NASA's international space activities.

3. On the evolution of NASA's international activities, see Logsdon, gen. ed., *Exploring the Unknown, Volume II*, Chapter One. Other related sources include Arnold Frutkin, *International Cooperation in Space* (Englewood Cliffs, NJ: Prentice-Hall, 1965), pp. 3–165; John M. Logsdon, "U.S.-European Cooperation in Space: A 25-Year Perspective," *Science*, Vol. 223, January 6, 1984, pp. 11–16; Kenneth Pedersen, "The Global Context: Changes and Challenges," in Molly McCauley, ed., *Economics and Technology in U.S. Space Policy* (Washington, DC: National Academy of Engineering, 1987; and Task Force on International Relations in Space, NASA Advisory Council, *International Space Policy for the 1990s and Beyond*, October 12, 1987.

4. National Aeronautics and Space Act of 1958, Public Law 85–568, Sec. 102(c)(7).

5. Arnold Frutkin, "International Cooperation in Space," *Science*, Vol. 169, July 24, 1970, pp. 333–39.

6. For a discussion of the decision to proceed with the Space Shuttle, see John M.

Logsdon, "The Space Shuttle Program: A Policy Failure?," *Science*, Vol. 232, May 30, 1986, pp. 1099–1105.

7. See Arturo Russo, *Big Technology, Little Science*, European Space Agency (ESA) HSR-19 (Noordwijk, Neth.: ESA, August 1997), and Logsdon, "U.S.-European Cooperation in Space," for discussions of this dissatisfaction.

8. Memorandum from LI-15/Director of International Affairs to MFA-13/Director, Space Station Task Force, "Strategy for International Cooperation in Space Station Planning," undated but August 1982 (hereafter referred to as Pedersen Strategy Memorandum). Robert Freitag, who was intimately involved in developing U.S.-European cooperation in the post-Apollo period, suggests that "the major reason for not involving Europe in joint development of the Shuttle was complexity of management which would have been exacerbated by the differences in technology experience." Letter to author, November 17, 1989. Another reviewer of an earlier draft of this section noted that the Department of Defense (DOD) intervened to block the possibility of European development of the Tug when it became clear that many highly classified U.S. national security missions would be using the Shuttle and would require an orbital transfer capability. To DOD, the use of a non-U.S. system for such a purpose was not acceptable. See Lorenza Sebesta, *United States-European Cooperation in the Post-Apollo Programme*, ESA HSR-15 (Noordwijk, Neth.: ESA, February 1995), for a fuller account of the events summarized here.

9. For a history of the Spacelab project that stresses its international aspects, see Douglas B. Lord, *Spacelab: An International*

Success Story (Washington, DC: NASA SP-487, 1987). For a European perspective, see Lorenza Sebasta, *Spacelab in Context,* ESA HSR-21 (Noordwijk, Neth.: ESA, October 1997).

10. For a discussion of the evolution of European space activity, see John Krige and Arturo Russo, *Europe in Space, 1960–1973,* ESA SP-1172 (Noordwijk, Neth.: ESA, September 1994).

11. One of the readers of an early draft of this section has commented: "The scars on both sides left by that (Spacelab) experience were a major factor all during the [space station] negotiations and still color working relationships. Attitudes about whether Europe did or did not make shrewd agreements, or get its money's worth, vary from one senior official to another. Still, Europe's regional attitude about 'never again' really drove a lot of things about the sharing arrangements and the legal regime agreed for the station. The experience was put to good use by Europeans selling the program at home." Comments were transmitted in a letter from NASA Historian Sylvia Fries to the author, November 27, 1989.

12. See Sebasta, *Spacelab in Context,* and Russo, *Big Technology, Little Science,* for discussions of the European assessment of Spacelab cooperation.

13. The remainder of this section is based on, in addition to the sources cited, interviews with Robert Freitag, May 31, 1988; Kenneth Pedersen, June 15, 1989; James Beggs, February 12, 1989, and April 27, 1990; Hans Mark, December 10, 1988; Margaret Finarelli, June 13, 1989; Lyn Wigbels, June 15, 1989; Gil Rye, June 19, 1989; Philip Culbertson, April 4, 1990; John Hodge, May 4, 1990; and Luther Powell, March 26, 1990. The author has also profited from several conversations with Terence Finn and Robert Lottmann regarding issues discussed in this section and from reviewers' comments on an earlier draft of this section as transmitted by the previously cited Fries letter of November 27, 1989, as well as comments on the draft by Robert Freitag and Richard Barnes.

14. For a discussion of this evolution, see John Logsdon and George Butler, "Space Station and Space Platform Concepts: A Historical Review," in Ivan Bekey and Daniel Herman, eds., *Space Stations and Space Platforms—Concepts, Design, Infrastructure, and Uses* (New York: American Institute of Aeronautics and Astronautics, 1985). See also Howard E. McCurdy, *The Space Station Decision: Incremental Politics and Technological Choice* (Baltimore, MD: Johns Hopkins University Press, 1990), for a history of the space station concept and of space station–related decisions within the United States throughout the period up to 1984.

15. Hans Mark, *The Space Station: A Personal Journey* (Durham, NC: Duke University Press, 1987), p. 121. The Mark book is a fascinating, highly personal account of the events leading to the approval of the space station.

16. This monograph is not a comprehensive history of the process leading to a decision to proceed with the space station. For such an account, see Howard E. McCurdy, *Space Station Decision.*

17. One product of the attempts in the first month of the Reagan administration to cut the NASA budget that had a significant effect on the development of space station cooperation was the cancellation in February 1981 of the U.S. spacecraft planned to be part of a joint NASA-ESA International Solar Polar Mission (ISPM). This cancellation was decided on without consultation with ESA, and it was met with outcries from ESA and European scientists and diplomatic protests from several European countries. Coupled with the Spacelab experience, canceling the ISPM spacecraft raised serious doubts in Europe of whether the United States was a dependable cooperative partner.

18. Advanced Programs, Office of Space Transportation Systems, NASA Headquarters, "Proceedings of Space Station Planning Workshop Held at the Michoud Assembly Facility in New Orleans, Louisiana on November 18, 19, and 20, 1981," undated.

19. A primary forum for these discussions was a joint NASA-ESA study group chaired by Ivan Bekey and Robert Freitag of NASA's Office of Manned Space Flight and Jacques Collet of ESA's Space Transportation Directorate. According to Freitag, James Beggs was fully aware of these discussions as he thought about international involvement in his space station initiative. Freitag letter to author, November 17, 1989.

20. "Proceedings of Space Station Planning Workshop."

21. Interview with Kenneth Pedersen.

22. Pedersen's remarks appear in Mireille Gerard and Pamela Edwards, eds., *Space Station: Policy, Planning, and Utilization* (New York: American Institute of Aeronautics and Astronautics, 1983), p. 116.

23. Interview with Kenneth Pedersen. On changes needed in NASA's approach to international space cooperation, see Kenneth Pedersen, "The Changing Face of International Space Cooperation," *Space Policy*, May 1986, pp. 123–30, and Kenneth Pedersen, "The Global Context."

24. "Space Station Briefing to Headquarters Officials," May 25, 1982.

25. Memorandum from MFA-13/John Hodge to LI-15/Kenneth Pedersen, "Strategy for International Cooperation in Space Station Planning," July 30, 1982.

26. Pedersen Strategy Memorandum, pp. 8–12.

27. *Ibid.*, p. 12. For an overview of the technology transfer controversy, see National Academy of Sciences, *Balancing the National Interest: U.S. National Security Export Controls and Global Economic Competition* (Washington, DC: National Academy Press, 1987). This study contains an extensive bibliography on export control and technology transfer issues.

28. Pedersen Strategy Memorandum, p. 12.

29. See McCurdy, *Space Station Decision*, Chapter 9, for a discussion of the task force approach.

30. It is interesting to note that Japan created a Space Station Task Force before the United States did; this was one indication of how eager the Japanese were to participate in the program.

31. Kenneth Pedersen, "Note for John Hodge—Space Station Efforts in Japan," May 5, 1982.

32. "Meeting of NASA Administrator and ESA Director General, June 17, 1982, ESA Head Office," Annex 3, ESA/NASA Space Station Planning Coordination.

33. *Ibid.*

34. Pedersen Strategy Memorandum, p. 3.

35. Interview with Robert Freitag.

36. Lyn D. Wigbels, Memorandum for the Record, "Space Station International Orientation Briefing," December 3, 1982.

37. Interviews with Kenneth Pedersen, Peggy Finarelli, and Robert Freitag and conversations with Robert Lottmann.

38. Telex from R.J. Barnes to Pedersen (LI-15), "European Participation in Space Station Study Kickoff Meetings—Sept. 13–15," undated. Among the major European countries, it was France that took the lead in pushing for an independent European approach to space; for all ESA members, the memory of the 1981 withdrawal of the U.S. spacecraft from the cooperative ISPM was still very much fresh.

39. Gerard and Edwards, eds., *Space Station*, p. 4. Particularly galling to space station partners throughout the period covered by this study, and even afterward, were the U.S. insistence of coupling its goal of leadership with invitations to others to cooperate (presumably as followers) and the explicit acknowledgment that one reason for inviting cooperation was to divert foreign resources from programs that might be competitive with the United States.

40. *Ibid.*, p. 117.

41. *Ibid.*, p. 122.

42. See McCurdy's *Space Station Decision* for a detailed discussion of SIG (Space) activity on the space station during 1982 and 1983.

43. Douglas R. Norton, "Technology Transfer and Space Station Planning," September 13, 1982. Norton was the person within NASA's Office of International Affairs who was responsible for interactions with the export control community.

44. NASA Space Station/Mission Analysis Studies, "Proposed Talking Points for DOD," undated but probably November 1982.

45. *Ibid.* For a review of various perspectives on the need to increase protection of technical information from unwanted transfer, see Harold Relyear, ed., *Striking a Balance: National, Security and Scientific Freedom* (Washington, DC: American Association for the Advancement of Science, 1985).

46. NASA Space Station/Technology Transfer, "Proposed Talking Points for Schneider-Meeting (Nov. 3, 1982)."

47. Interviews with Kenneth Pedersen and Peggy Finarelli.

48. Letter from Kenneth Pedersen to Gordon Woodcock of Boeing Aerospace, December 14, 1982. Similar letters were sent to the other seven mission requirements contractors.

49. Interviews with John Hodge, Phil Culbertson, and Robert Freitag.

50. Interviews with Pedersen and Finarelli.

51. Letter from James Beggs to Edwin Meese III, May 21, 1982.

52. Reagan's speech is reprinted in Mark, *Space Station*, p. 249. The Mark book contains a description of this early attempt to gain presidential endorsement for the space station.

53. *Ibid.*, p. 247.

54. *Ibid.*, p. 162.

55. "SIG Working Group Talking Paper," November 30, 1982.

56. Memorandum from MFA-13/Executive Secretary, SIG-Space Station Working Group, to SIG-Space Station Working Group Members, "Record of SIG-Space Station Working Group Meeting, November 22, 1982."

57. Mark, *Space Station*, p. 251.

58. "Space Station Presentation to the President," April 7, 1983.

59. McCurdy, *Space Station Decision*, Chapter 17. McCurdy's book contains a detailed account of the steps in the summer and fall of 1983 leading to the presidential decision to approve the space station.

60. Mark, *Space Station*, pp. 164–65, and interviews with James Beggs and Hans Mark.

61. *Ibid.*, pp. 178–80.

62. Letter from James Beggs to Robert C. McFarlane, October 31, 1983.

63. Memorandum from LI-15/Director of International Affairs to FA-13/Deputy Director, Space Station Task Force, "Policy on International Involvement in a Space Station," July 20, 1983.

64. Interviews with James Beggs, Gil Rye, and Peggy Finarelli.

65. Letter from the President of the United States, January 25, 1984.

66. The draft terms of reference as circulated by Gil Rye on February 6, 1984, said that NASA "would take the lead" in preparing this report and that "other U.S. Government agencies will be invited to participate in the report's preparation." The final terms of reference issued on February 25 said that "NASA will take the report" and specified that the Department of Commerce, the Department of Defense, and the Director of Central Intelligence would also be involved in its preparation. Letter from Robert McFarlane to James Beggs, February 25, 1984.

67. *Ibid.*

68. Interview with Peggy Finarelli and letter to author from Thomas Niles, May 1, 1990. See the subsection in this monograph titled "The Partners Accept the Invitation" for more details on the link between space station cooperation and summit activities.

69. "Terms of Reference" attached by McFarlane to Beggs letter, February 25, 1984.

70. Interview with James Beggs.

71. Letter from James Beggs to James A. Baker III, February 19, 1984.

72. Memorandum from LI/Director of International Affairs to A/Administrator, "Potential Foreign Contributions to a U.S. Space Station," January 24, 1984.

73. Letter from James Beggs to George Shultz, March 16, 1984.

74. *Ibid.* and interviews with James Beggs, Kenneth Pedersen, and Peggy Finarelli.

75. Letter from James Beggs to Kenneth Baker, U.K. Minister of Trade and Industry, April 6, 1984. Similar letters were sent to top-ranking individuals in other countries that Beggs had visited.

76. Letter from Robert C. McFarlane to James Beggs, February 25, 1984.

77. Letter from Thomas Niles to the author, May 1, 1990.

78. Memorandum from LI/Director of International Affairs to Administrator, "Space Station at the London Economic Summit," February 28, 1984.

79. Letter from James Beggs to Honorable H. Allan Wallis, March 29, 1984.

80. Letter from James Beggs to U.K. Minister of Trade and Industry Kenneth Baker, April 6, 1984. A similar letter was sent to other top officials met during the trip.

81. *Ibid.*

82. Letter from Beggs to Shultz, March 16, 1984.

83. One potential partner had already, and enthusiastically, made the decision to participate. Even before NASA Administrator Beggs arrived in Rome, Italian President Craxi wrote to President Reagan, telling him that a March 5 meeting of the Italian cabinet had decided that "Italy . . . is quite ready to study the terms of significant cooperation." Letter from B. Craxi to Ronald Reagan, March 6, 1984.

84. *Economic Summit Communiqué*, June 9, 1984.

85. Commenting on an earlier draft of this section, former NASA European representative Richard Barnes remarked that "the watering down of the language on Space Station [in the Summit communiqué] was a direct result of last minute intervention by Jacques Attali, the French Sherpa and President Mitterrand's right hand man at the Elysee. This was the first, but by no means the last, example of behind-the-scenes maneuvering by the French aimed at impeding the dialogue on cooperation." Personal communication to the author, April 5, 1991.

86. Hubert Bortzmayer, "Space Station Poses Dilemma for Europe," *Aerospace America*, February 1984, p. 26.

87. In addition to the specific sources cited below, this account of European decision-making is based on interviews with Jean Arets, Head of International Affairs, ESA, and Gabriel Laferranderie, Legal Counsel,. ESA, respectively, on January 6, 1989, and February 16, 1990, Jacques Collet, Space Transportation Systems Directorate, ESA, on January 7, 1989, and Roger Bonnet, Director for Science, ESA, on January 7, 1989, as well as on several conversations with Reimar Luest, Director General, ESA, and George van Reeth, Director for Administration, ESA. The author also was helped by conversations with officials of the French space agency (CNES), including Daniel Sacotte, Director for International and Industrial Affairs, Isaac Revah, Director of Programs, and Alain Dupas, Long-Range Program Planning, as well as with Hans Hoffman of MBB/ERNO, Gottfreid Greger of the German Ministry for Research and Technology, and James Zimmerman, NASA European Representative. Finally, the author benefited from comments on earlier drafts of this section by Richard Barnes, former NASA European Representative, Peggy Finarelli, and Robert Freitag.

88. Resolution ESA/C/X/Res. 2, October 8, 1976.

89. Kenneth Pedersen, "Memorandum for the Record," February 19, 1982.

90. Jesco von Puttkamer, "Memorandum for the Record: Results of NASA–ESA Advanced Programs Coordination Meeting, February 10–11, 1982, NASA Headquarters," February 11, 1982.

91. Resolution Concerning the Space Transportation Systems Long-Term Preparatory Programme," ESA/C/LIV/ Res. 1, June 22, 1982.

92. Annex A to "Declaration Concerning a Preparatory Programme for Long-Term Space Transportation Systems," ESA/C/LV/Dec. (Final), October 6, 1982, updated on February 23, 1983.

93. Memorandum to Director of International Affairs from NASA European Representative, "Highlights of December 8–9 ESA Council Meeting," December 23, 1982.

94. Translation of *Le Figaro* article (undated) by NASA European Representative, February 9, 1983.

95. Jeffrey Lenorovitz, "Germany, Italy Propose Space Station, *Aviation Week and Space Technology*, February 20, 1984, pp. 55–56. Quote is on p. 55.

96. For background on French planning, see Jeffrey Lenorovitz, "French Plan Unmanned Space Station," *Aviation Week and Space Technology,* August 3, 1981, pp. 49–51, and David Dickson, "France Pushes Europe Toward Manned Space Flight," *Science,* January 17, 1986, pp. 209–10. The quote is from a NASA memorandum from James Morrison, Deputy Director of International Affairs, to Director of Space Station Task Force, August 16, 1982.

97. ESA efforts are divided between mandatory programs, to which member states must contribute according to a preset formula, and optional programs, to which member states contribute based on their political and industrial interest. Scientific programs are part of the mandatory portion; most large system development activities such as Spacelab and Ariane were optional programs.

98. ESA Resolution ESA/C/LXIV/Res. 4 (Final), June 28, 1984.

99. European Space Agency, "Outline of a Long-Term European Space Plan," ESA/C (84) 46, Rev. 1, November 21, 1984, pp. 2–3.

100. *Ibid.,* pp. 16, 17, 22.

101. Erik Quistgaard had been replaced in mid-1984 by Reimar Luest, a German scientist/administrator. It was Luest who took the lead for ESA in rallying member-state support behind the agency's long-range plan.

102. Eurospace, *Towards a European Long-Term Space Programme,* 1985.

103. In an attempt to gain the support of some in the space science community for the space station, NASA had included as part of its station program several large automated Earth observation platforms in a polar orbit quite separate from the core station complex. The logic for this inclusion was somewhat tenuous; scientists were told by NASA managers that the chance of getting approval for these observation platforms would be enhanced by associating them with the politically strong space station. The reality turned out to be somewhat different, and the platforms were separated from the station program in the late 1980s.

104. This discussion is based on an interview with Peggy Finarelli, a NASA memorandum from Deputy Director, Policy and Plans Office, Office of the Space Station, to Director, Policy and Plans Office, Office of the Space Station, "London Trip," November 5, 1984, and a letter from Jack Leeming, Department of Trade and Industry, to John Hodge, NASA, August 16, 1984.

105. Summary of meeting attached to a memorandum from Robert Freitag, "UK Policy on Space Station," November 2, 1984.

106. Memorandum from Deputy Director to Director, "London Trip," November 5, 1984.

107. Dickson, "France Pushes Europe."

108. Frederic d'Allest, "A Space Policy for Europe," *Le Monde,* January 29, 1985, as translated by NASA European Representative Richard Barnes.

109. Memorandum from Director of International Affairs to Administrator, "Visit of Dr. Hans Riesenhuber," October 14, 1983.

110. Letter from Hans Riesenhuber to James Beggs, October 27, 1983, as translated by the German embassy in Washington, DC.

111. Translation of the Kohl letter (undated) is attached to a memorandum from Director of International Affairs to Administrator, "Reagan Letter to Kohl," January 20, 1984.

112. Personal communication from Richard Barnes, April 5, 1991.

113. Note to Otho Eskin, Department of State, from Margaret Finarelli, NASA, "Kohl Letter to the President," February 6, 1985.

114. Cable from Ambassador Arthur Burns to Secretary of State, "Research and Finance Ministers Agree on Space Station Financing: Final Roadblock to German Participation Removed," January 1985.

115. Cable from Ambassador Arthur Burns to Secretary of State, "Press Release on German Participation in U.S. Manned Space Station," January 1985.

116. Cable from Ambassador Arthur Burns to Secretary of State, "FRG Cabinet Gives Go-Ahead to Space Station and Ariane 5 Program," January 1985.

117. Council, European Space Agency, "Resolution on the Long-Term European Space Plan," ESA/C-M/LXVII/Res. 1 (Final).

118. Ibid.

119. ESA Council, "Resolution on Participation in the Space Station Programme," ESA/C-M/LXVII/Res. 2 (Final), January 31, 1985.

120. For a discussion of this history, see Logsdon, "U.S.-European Cooperation in Space Science."

121. In addition to the sources cited below, this account of Japanese consideration of the U.S. invitation to participate in the space station program is based on interviews with Kaoru Mamiya and Masahiro Kawasaki of the Science and Technology Agency and Masatoshi Saito and Masafumi Miyazawa of NASDA, May 15, 1991.

122. See the earlier section titled "Origins of the U.S. Invitation to Cooperate" for an account of how the United States, during the 1970–1972 period, changed the conditions under which foreign participation in the post-Apollo program was acceptable. These changes applied to Japan as well as to Europe.

123. "Japan's Space Program: A National Priority," interview with Nobuytki Arino, Executive Managing Director, TRW Overseas, Aerospace America, March 1985, p. 65.

124. National Space Development Agency, NASDA 185-186, September 1985, p. 8.

125. Space Activities Commission, "Outline of Japan's Space Development Policy," revised on February 23, 1984 (unofficial translation), p. 7.

126. Andrew Pollack, "Japan's Space Race Struggle," New York Times, August 24, 1984, pp. D1, D5. For an overview of U.S.-Japanese space relations, see John M. Logsdon, "U.S.-Japan Space Relations at the Crossroads," Science, Vol. 225, January 31, 1992, pp. 294–300 and John M. Logsdon, Learning from the Leader: The Early Years of Japanese-U.S. Space Relations (Washington, DC: Space Policy Institute, George Washington University, 1998).

127. Ibid.

128. Memorandum from Director of International Affairs to Director, Space Station Task Force, "Japanese Space Station Planning," July 15, 1982.

129. Telegram from A. Kubozono, NASDA, to K. S. Pedersen, NASA, "NASDA Space Station Task Force," July 27, 1982.

130. NASA, "Space Station Planning Activities in Japan," December 8, 1982.

131. Mitsubishi Group, "Proposed Japanese Participation in the U.S. Space Station Program," presented to Japanese government, September 1982.

132. Letter from James Beggs to George Shultz, January 4, 1983.

133. Philip Culbertson and Terence Finn, "Visit to Japan—March 6–11, 1983," p. 2.

134. The subcommittee's conclusions were reported by Masatoshi Saito of NASDA at an American Institute of Aeronautics and Astronautics Symposium in July 1983. See Gerard and Edwards, eds., Space Station, pp. 120–22.

135. Informal U.S. Embassy translation of Nihon Keizai article of October 17, 1983, contained in cable from Ambassador Mansfield to Secretary of State, October 1983.

136. Translation of article that appeared in Nihon Keizai Shimbun, November 21, 1984, p. 3 (JPRS-JST-85-018-L). MITI had been trying to get involved in various areas of space activity that NASDA viewed as its "turf" since 1980.

137. Ibid.

138. Letter from Philip Culbertson to Makoto Miura, December 6, 1984.

139. Cable from Makoto Miura to Philip Culbertson, "Japanese Space Station Phase B Budget," December 28, 1984.

140. Letter from Tadahiro Sekimoto to James Beggs, March 6, 1985.

141. Ad Hoc Committee on Space Station Program, Space Activities Commission, "Basic Plan for Japanese Participation in the Space Station Program" (Final Report—unofficial translation), April 10, 1985, pp. 14–16.

142. In addition to the sources cited below, this account of Canadian consideration of the U.S. invitation to participate in the space station program is based on interviews with Mac Evans and Karl Doetsch, respectively, Canadian Space Agency, on June 10, 1989, and March 13, 1990, and with William Cockburn, then Counsellor for Science and Technology, Embassy of Canada, on August 10, 1988.

143. "Canadian Astronaut Program Announced," Press Release, Ministry of State for Economic Development, Science and Technology, June 8, 1983.

144. Doetsch's remarks are in Gerard and Edward, eds., *Space Station*, p. 118.

145. National Research Council of Canada, "Canadian Participation in Space Station, in John Kirton, ed., *Canada, the United States, and Space* (Toronto: Canadian Institute of International Affairs, 1986), p.110.

146. *Ibid.* and interview with William Cockburn.

147. Doetsch in Gerard and Edwards, eds., *Space Station*, p. 118.

148. Memorandum from Kenneth Pedersen, "Results of Space Station Trip," March 21, 1984.

149. Karl Doetsch, "Space Station-Canadian Considerations for Participation," paper enclosed with a letter from Doetsch to P.E. Culbertson, NASA, December 20, 1984.

150. Interview with William Cockburn.

151. Cable from Ambassador Robinson to Secretary of State, "Canadian Manned Space Station Considerations," January 1985.

152. Cable from Secretary of State Shultz to American Embassy, Ottawa, "Canadian Contribution to the Manned Space Station," February 1985.

153. *Canada's Interim Space Plan*, 1985–1986, reprinted in Kirton, *Canada, the United States, and Space*, p. 101.

154. Summary, "The Quebec Summit," March 18, 1985.

155. Quoted in Kirton, *Canada, the United States, and Space*, p. 100.

156. See, however, John M. Logsdon, "International Cooperation in the Space Station Programme: Assessing the Experience to Date," *Space Policy*, February 1991, pp. 35–46.